Service Agreements for SMB Consultants

A Quick-Start Guide to Managed Services

Revised Edition

Karl W. Palachuk

Author, The Network Documentation Workbook
and Managed Services in a Month

Great Little Book Publishing Co., Inc.
Sacramento, CA

www.greatlittlebook.com

Great Little Book Publishing Co., Inc.
Sacramento, CA

Service Agreements for SMB Consultants — A Quick-Start Guide to Managed Services – Revised Edition

www.greatlittlebook.com

ISBN # 978-1-942115-49-6

Service Agreements for SMB Consultants

A Quick-Start Guide to Managed Services

Revised Edition

Karl W. Palachuk

Table of Contents

This book is dedicated to my daughter Victoria, the greatest thing that ever happened to me and the love of my life.

— Dad

Your Downloadable Content

This book includes a few additional downloads what you will find very helpful. These include Word and Excel files, and a few other goodies.

If you purchased this book from SMB Books or Great Little Book, you should have received a download link when your purchase was completed.

If you lost that or purchased from Amazon or another reseller, you can register at **www.smbbooks.com**.

Please have your purchase receipt ready to register. You'll need the Order ID. If your purchase somewhere else, you'll need to forward proof of purchase to us.

Your feedback is always welcome.

The Plan of the Book

This book has four basic sections. Chapters One and Two address who you are and who your clients are. Chapter One covers various ways to formulate your business. What worked yesterday may not work tomorrow. Chapter Two discusses the various relationships that can exist between you and your clients. More specifically, it covers the various ways you can sell your services.

The second section, Chapters Three through Six, covers the nuts and bolts of building a service agreement. Chapter Three discusses the "parts" that make up a contract. Think of this like a set of building blocks. You don't have to use all the parts, and the way you put them together will define any specific project.

Chapters Four, Five, and Six are "boilerplate" agreements that can be used as a starting place for writing your own agreements. Chapter Four covers basic by-the-hour work without a "contract." In other words, before you do break-fix work, whip out this credit agreement and get a signature. Chapter Five is a basic agreement for hourly services (whether billed by the hour or via blocks of time). And Chapter Six is a sample agreement for flat fee or managed services.

The third section covers a bit of the management side of working with contracts or service agreements. Chapter Seven discusses using lawyers, accountants, and other professionals to help you manage your success. Chapter Eight covers keeping track of your service agreements. After all, service agreements expire, rates increase, and the world keeps on turning.

Finally, the last section consists of a Conclusion and some resources to help you along your way.

A Note About KPEnterprises

For sixteen years I owned and operated KPEnterprises Business Consulting, which has been the model for my experiences and writing over the last decade. But people and businesses evolve.

KPEnterprises was closed down at the end of 2011 and is now simply a brand name underneath Great Little Book Publishing Co., Inc. I am spending most of my time writing, consulting, and training technology consultants. KPEnterprises became America's Tech Support (ATS), and I worked there for a few years.

I worked as the Senior Systems Engineer for ATS. I was responsible for strategic planning, some sales, some project management, and some network migrations.

Then, in 2014 I started consulting on my own again under the Small Biz Thoughts brand. At the end of 2016 that business had grown and was taking too much time. So I sold it as well. I am now a coach and do backup tech support for the new owner.

This mix works great for me. I get to play with new technology. I get to interact with clients. I get to keep my fingers in the support side of the business.

So . . . when I refer to KPEnterprises or Small Biz Thoughts in this book, I am either referring to the company I owned for sixteen years, or ATS, or my current arrangement. All of these operated on the principles and guidelines discussed in this book.

I hope you find this little book useful. I welcome your feedback. Send me an email at **karlp@smallbizthoughts.com**. Let me know how you're doing.

Who Should Read This Book?

This book is intended for SMB (small and medium business) I.T. consultants who are creating or revising their service agreements.

(Note: If you think you are a "computer consultant" you really need this book. It's a great introduction to the concept of moving from the computer consultant to the professional I.T. consultant. Our modern term for that is Managed Service Provider.)

While this book is most useful for those who have never had a service agreement, it is also useful for more experienced companies that are revising their offerings or otherwise rethinking their business.

Many companies have been "break/fix" since day one. Now, as they look around, it seems like everyone's talking about "managed services" (whatever that means). If you're in this position, you will find this book very helpful. Because there's a strong focus on best practices and how you define your relationship with your clients, everyone should find something of value here.

So, if you're new to the business, this book will be a great big step up in the area of managing client relationships. If you've been in business a while but not used service agreements before, you'll enjoy the discussion of how your client relationships are changing as well as the boilerplate service agreement text.

Finally, consultants who have been using service agreements will benefit from a different perspective, as well as some of the business management discussions. In addition to simple service agreements, the book covers how you define your organization. Even if you're successful in the managed services area, you'll enjoy the discussions of how you formulate your business and manage your agreements.

You should have a lawyer review every agreement or contract you sign.
www.smallbizthoughts.com

How to Use This Book

As with any good book, you should read this book with a pen or pencil in hand. Mark up the borders. Make notes to yourself. Put in a few comments where you agree or disagree – both of these are important.

Don't gloss over that point. Your opinion and your input into this process are far more important than mine. My role is to give you some thoughts on running your business and give you a place to start with service agreements.

One of the things I've learned from traveling around the world and talking to consultants is that their businesses are personal and unique. No two are exactly alike. At the same time, it has been extremely valuable to exchange ideas on how we run our businesses.

Please do not toss the book onto a shelf, download the contract text, and start using the service agreement language as your own! Aside from all the legal warnings, you'll be skipping the important information you need to make it all make sense.

Don't skip the "installation notes." Maybe I should have titled the book

README.TXT.

Chapters One and Two, and Chapters Eight and Nine, are the business discussion. You'll get the most out of the business discussion by keeping an open mind about how you do things. At the same time, don't reject practices that have been working for you because you read something in a book! I don't live in your city, compete in your environment, or own your company.

You should have a lawyer review every agreement or contract you sign.
www.smallbizthoughts.com

The middle chapters are the service agreement language. Again, mark this up and make notes. In particular, I discuss options and the reasons for making some of the choices of language. Consider this discussion before you blindly copy something from the downloads.

Having convinced you to put off the download text until last, let me also encourage you to open the download zip file and look at materials other than the service agreements. You'll find a few Excel workbooks with calculations that may be useful. You'll also find sample documents for keeping track of your agreements. And, finally, I've also thrown in some links to the IRS as well as some forms that you may find useful.

The bottom line – use the book! You didn't buy a collection of legal samples, you bought a book that comes with a collection of legal samples.

Legal Disclaimer

I'm not a lawyer. Nothing in this book is to be considered legal advice. You are responsible for all actions you take and any decisions you make.

The samples in this book have been reviewed by a lawyer in California. That's about as useful as a VD test – it's good for about twenty minutes. Laws change. Judges make rulings that affect interpretations. The law is a living thing.

In addition to that, please be aware that some clauses do not work well together. Other clauses are not binding in all localities. So the bottom line is:

You are responsible for having your lawyer review any agreements or contracts you write or sign.

If you take boilerplate text out of this book, you may very well end up with an un-enforceable contract.

Neither Karl W. Palachuk nor Great Little Book Publishing Co., Inc. can be held responsible for actions you take or agreements you create as a result of the materials found in this book.

'Nuf Said.

Section I
Defining Yourself and Your Relationships

Chapter 1
Define Yourself

Chapter 2
Define Client Relationships

"Life is too short to be unhappy in business."
— George L. Brown

Chapter One – Define Yourself

Introduction

It may sound overly dramatic, but the winds of change are coming to SMB Consultants. A storm is gathering around us. We can ignore the change, we can fight the change, but we can't stop the change. There is one more alternative – we can make our businesses evolve to take advantage of the change.

Technicians who support Small and Medium-sized Businesses (SMB) are virtually all small businesses. Even the larger consulting houses tend to be in the range of ten to twenty consultants. A consulting company with fifty or more technicians is very rare.

After roughly thirty years of quiet anonymity, our businesses are now the focus of three major movements. After watching one industry after another go through the process of growth and consolidation, it's our turn.

During the Dot Com Bubble we put up with technicians who wanted $70,000 salaries for entry level jobs. After the bubble burst, some of those unemployed technicians competed against us at $25 per hour. But the forces of change were not always as visible.

After the Tech Wreck of the early 2000's, three movements emerged that are changing our business whether we like it or not. These movements are loosely related to one another.

The first movement is the most mature – Managed Services. I define managed services as Technical Support delivered under a service agreement that provides specified rates and guarantees the consultant a specific minimum income.

It can mean flat rate pricing to some. To others it means remote administration or remote monitoring. It can be anything from a home grown collection of tools and procedures to a complete software and services package.

The second movement is the growth of the franchised "geek" industry. The Best Buy Geek Squad has the mindshare in this business right now. They're kind of the Kleenex of mobile computer technicians. Geeks on Call, Computer Troubleshooters, and several other companies are fighting for the P.R. success of the Geek Squad.

These are still early days for the franchised "geek" industry. I don't know if the current service model and pricing model are sustainable. But you can be sure of this – they're not going away. They may need to be rebuilt and re-constituted. There may be an industry shakedown. If Best Buy is successful, others will follow. But some form of franchised or licensed/branded technical service will survive.

You can also be sure of this: Branded technical support will move from home users to small business users—and eventually to medium size businesses. Best Buy is already opening Geek Squads that focus on small businesses.

Which leads us to the third major movement – major national companies are beginning to compete with us, the small consultants. The most notable entry into this field is Dell Computers. They like to call their offering "Managed Services." In fact, they're combining remote administration, help desk support, and local I.T. consultants to provide a one-stop-shopping package to small businesses.

The Dell Managed Services offering is very immature. It has the most visibility among SMB consultants because many of them consider Dell a "partner" and that partner is now competing with them. As with the services mentioned above, the Dell model may need to be restructured to be profitable. At the same time, others will join this fray.

In addition to those three movements, there's a rising fever for Mergers and Acquisition (M&A) in the small end of the SMB consulting world. Several organizations have grown up specializing in helping buyers and sellers of IT consulting firms find each other and strike a deal.

The M&A activity started to head up in the mid-2000's simply because our industry was reaching a certain maturity. When the recession hit in late 2008, many small companies either ceased to exist or sold off their lists in order to get some extra cash before they went out of business. Today – ten years later – we're seeing a lot of people selling because they want to retire.

Is all of this depressing? No! What a wonderful, busy, exciting time to be in the industry! This is exactly the environment in which entrepreneurs flourish! Companies like Dell and Best Buy could take years to figure out what they're doing. And we, the "little guys," have the clear advantage. When they zig, we can zag. We can change directions, fill in the gaps, clean up the messes, and make money in the process.

And best of all, this is the perfect time for us — for SMB consultants — to redefine our businesses.

That's where this book comes in. Your contracts or service agreements are the very definition of the services you offer and the prices you charge. They are the formal definition of your relationship with your clients. There is always the personal side, of course, and that's what keeps small businesses in business. But the basic description of how you operate is defined by your service contracts.

As we look ahead to the year 2025, it is the job of the business owner to anticipate what the market will look like and to position the business to be successful in that market. Here's what we know:

➤ **Managed services will be everywhere.**

As a result, a certain piece of our business will be "commoditized." That is, there will be a large number of chores that anyone can do and they will be farmed out to the lowest bidder.

➤ **Branded technical support will be everywhere.**

Again, that means that certain services will become commodities. It also means that you have opportunities to join a franchise, buy into a licensed brand, join a technical group, or start your own organization.

➤ **One-Stop-Shopping from national companies will be everywhere.**

This is really a combination of the previous two that is implemented by a large corporation. Your opportunity may be to take part in the process, use them where you can, and compete where they can't.

➤ **Due to M&A activity, there will be some very large players in our business who are very professional.**

For the most part, the large players such as Staples, Ingram Micro, and Dell have had no real ability to compete with us. They have size but no real experience delivering "boots on the ground" tech support to small business. The opposite is true of the large MSPs (Managed Service Providers) made up from smaller companies with lots of experience.

➤ **TCP/IP and IoT bring you more opportunity to succeed**

than anything that's emerged in the last two decades.
Virtually every evolving technology is using TCP/IP and Internetworking to deliver services. As IoT (Internet of Things) evolves, it will create massive networks that need to be managed. Suddenly, we have an almost unlimited set of opportunities in front of us.

➤ **None of these can lead to the demise of small I.T. consulting firms.**

One piece of the market will consist of commodity services. One piece will consist of clients who will never use managed services or branded support. And one piece will always consist of specialty services that cannot be supplied by the "competition" mentioned above.

This book has a very simple goal – to provide SMB consultants with a solid introduction to support agreements. My goal is to make this book immediately useful. Notice that I didn't call this "The Big Fat Book of . . ." or "Everything You Need to Know . . . ." This book is not the "Bible" of consulting agreements.

This book is a Quick Start Guide. It covers the basic information you need to write a good service agreement in the modern era. If nothing else, it should give you a place to start thinking about how you will formulate your business as these winds of change blow across the landscape of SMB consulting.

I believe these changes are coming—in fact they're already here. Five years from now, almost every aspect of small biz I.T. consulting will be different. Now is the time to figure out where your business is going and how to get there. Every SMB consultant without a business plan needs one. You need to define who you are and where you're going. This book provides one tiny piece of that puzzle.

Define Yourself

The first step in defining your relationship with your client is to define who you are. Are you a sole proprietor? Are you a corporation? An LLC? Something else?

There are really only a few ways to define yourself. And, really, the two default methods are sole proprietor and S-Corp. We'll discuss why below. Pretty much everyone starts out as a sole proprietor. Once you get to a certain size, or a certain salary, then S-Corp makes sense.

To your clients, only two things matter:

> "To whom do I make the check out?"

and

> "Do I need to send 1099s at the end of the year?"

Other than that, the only time you really need to pay attention to how your business is legally defined is in your service agreements. Of particular importance, as you'll see in the "boilerplate" sections, is the question of getting out of the way of Uncle Sam.

Note to non-U.S. readers: This section is U.S.-centric. I have enough work trying to keep up on U.S. tax laws without trying to address the U.K., Canada, Australia, India, and the rest of the world. I think the general issues are the same in other countries, but the specific laws will vary dramatically.

Use what you find here as a common-sense discussion, but absolutely find the right professional assistance for your country.

Forming Your Consultancy

I'm sure you're a very nice person. But if I were to engage in business with you, I'd want to make sure that your personal "stuff" doesn't get all mixed up with my personal "stuff." If you've ever been a landlord, you know what I mean. In business, we need to separate our personal financial life from our professional financial life.

This section covers a brief overview of the various forms your business might take. If you are an S-Corp, or have recently discussed all this with your accountant or enrolled agent, then you might skip to the next section (entitled "Important Safety Tip: Don't Mess With the IRS"). If you're just starting out, are a sole proprietor, LLC, or some other type of entity, this section is probably worth reading.

In general, I think that sole proprietors should re-address the question of incorporation every few years. In particular, if you are successful, then revisiting the benefits is in order.

Another area affected by your business form is your status as an employer. When you hire people, then you have a whole new world of taxes and forms and filings. And now you have your personal stuff exposed to your employees' personal stuff.

Important Safety Tip: Don't Mess With the IRS

People go bankrupt when they ignore the IRS's instructions. This is not for you.

You need to be very careful to make sure that your service agreements are written so that you cannot be considered an employee of your clients. You can say "I'm not." But that only goes so far. The IRS rules change all the time, but two issues will always affect independent consultants.

First, the question of whether the client should be withholding taxes from your "paycheck." Second, the question of whether you can take a home office deduction.

On both questions, I'm deep into the "play it safe" school of thought.

We're not going to look at the home office deduction very closely as it's not the focus of this book. But you need to talk to your CPA or Enrolled Agent and take his advice. As for your Uncle, go to http://www.irs.gov and do a search of "home office."

On the question of employee versus independent contractor, we have several items in the service agreements chapters that are intended to define this relationship. Again, to find out the latest rules and regs, go to http:// www.irs.gov and search for "Employee or Independent Contractor?" You might also look at IRS form SS-8, "Determination of Worker Status for Purposes of Federal Employment Taxes and Income Tax Withholding."

Just so you know, your Uncle doesn't care how you see yourself or your relationship with the client. Uncle Sam has some very specific ideas about whether or not you're an employee. The general rule is: If the client controls what will be done and how it will be done, then the client is your employer.

Of course, this is broken down into specific discussions. For example, if the client tells you where to be, when to be there, how to get the job done, what tools to use, who to hire, where to buy supplies, and in what sequence work must be performed, then you are an employee.

Similarly, if the client pays for training, you might be an employee.

You might look at the IRS publications and find yourself on a borderline for one or more regulations. That's where contracts or service agreements come in. The IRS specifically states that one of the criteria for defining the relationship is to specify that relationship in a contract.

So, you see why it's important to make sure you never do any work without some kind of agreement. Even the "Credit Agreement" we present covers the basics of the relationship. If anyone balks at signing this, you can tell them two things. First, you won't work without it. And, second, one of the goals is to define your relationship so they don't have to withhold taxes and put you on their payroll.

As for the longer service agreements, just make sure you cover all the major points the IRS might want to look at. State very specifically that you, the consultant, will determine what needs to be done, what tools are necessary, the order in which work will be performed, etc. Also state very plainly that you are a contractor and not an employee, and that you'll pay your own taxes.

And that's the key to success – who's paying all those taxes? After all, the IRS does not exist to make the world a better place (which is good, because they suck at it). The IRS exists to collect various kinds of taxes. You don't have to specify that you'll pay your Federal taxes, and your state taxes, and your unemployment taxes, etc. But it goes a long ways to have a signed agreement that both you and the client know that you are paying your taxes.

The bottom line: Find out what the current rules are and rely on professionals for advice. The tax business is just as fast-paced as the computer business. By the time something gets printed, it's probably out of date. Find a good Certified Public Accountant or Enrolled Agent to advise you on these matters. And make sure you have a lawyer review your service agreements!

If you are the kind of consultant who actually "goes to work" at a client's office, has a desk to sit at, and you perform most or all of your work there, you need to be particularly careful about these regulations. You also need to be extremely careful about taking a home office deduction.

Define Yourself: Sole Proprietor

Perhaps the "default configuration" for an independent contractor is to be a sole proprietor. That means that you are just you for tax purposes. Part of you runs a business, and that part of you has to fill out Schedule C on your income taxes.

Legally it means that you personally are doing business with each of your clients. There is no entity between you and your clients, such as a corporation. Unless your business name is your name, then you need to file a DBA ("Doing Business As") or Fictitious Business Name filing with some government agency.

For example, my business started out as a sole proprietorship. I was Karl W. Palachuk, DBA KPEnterprises. That allowed me to get a bank account under the name KPEnterprises. But it was still associated with *my* Social Security number. Clients could write checks to KPEnterprises or to Karl Palachuk. It didn't matter because we were one and the same.

You can grow very large and remain a sole proprietor. If you ask your clients you might find some surprisingly large companies that are really sole proprietorships. You can have employees, buy equipment, and depreciate assets. You can do 99% of all the things any other business does.

The advantage of being a sole proprietor is that it's easy. You just start selling services and goods. You pay your bills, you collect money. If you make a profit, that goes onto your personal tax return and you pay taxes.

If you have a professional do your taxes, it's certainly cheaper to do one extra Schedule C than to do a corporate tax return in addition to a personal tax return. So, there's an advantage there.

The disadvantages of a sole proprietorship (in my opinion) are all financial. First, your business is your personal life. Your "stuff" is mixed up with your clients' "stuff." In particular, if your client is also a sole proprietor, then their personal stuff is mixed with your personal stuff.

I know that bad things almost never happen. In more than twenty years in my business (seven of them as a sole proprietor), I never even had a small hint of a problem along these lines. Nobody came after my house, no one put a lien on the contents of my garage, etc. But it is a fact that a sole proprietorship is just you personally and you need to be aware of that.

You can limit your exposure to lawsuits by purchasing Errors and Omissions (Liability) insurance. My company has paid from about $400 per year up to $1,600 per year for a basic E&O policy. It depends on the number of employees, number of clients, and the nature of the work you provide. I believe lawsuits are rare, but you still need to take seriously the possibility that someone will sue your business.

A second disadvantage to a sole proprietorship is the self-employment tax. This falls into the category of "Don't get me started!" If you are an employee, roughly 7.5% of your wages goes to Social Security and Medicare, where it is immediately loaned to the Federal Government for research on wasteful spending habits of political hacks. But you're not alone. Your employer matches this amount. So, really, a number equivalent to about 15% of your income is flushed down the government toilet. (Note: this number changes a little from time to time.)

When you are self-employed (as a sole proprietor is), you get to pay both sides of this, which is fine. If you were a corporation, you would also pay both sides, but you the person and you the corporation would each pay 7.5%. But here's the bad news:

As a sole proprietor, all of your profit is considered your personal income. If you earn $60,000, you pay Social Security taxes on $60,000. If you make $90,000, you pay Social Security on all $90,000.

If you were an S-Corp, you'd pay yourself a salary and pay the Social Security only on your salary. The rest of the profit from the business would flow to you as "dividends." So, let's say you pay yourself that very reasonable $60,000 salary. You pay the Social Security and Medicare on $60,000.

But you don't pay the Social Security and Medicare on the remaining $30,000. Fifteen percent of $30,000 is $4,500! That's a chunk of change. You'd still pay your regular tax rate on that $30,000, but not the Social Security. Note that you don't pay Social Security taxes after a certain income level (currently around $127,000, but this goes up every year). And if you take a salary, you have to pay an additional .9% Medicare tax on earnings over $200,000.

Your numbers may be very different. Check out the IRS's "Tax Topic 751" for more info:

https://www.irs.gov/taxtopics/tc751.html

The point is simple, however. There's a lot of money at stake here and you should give some serious consideration to how your company is formed. Note that you cannot pay yourself a miserably low salary to avoid Social Security taxes. The basic rule for the IRS is pretty simple. If you do something just to avoid taxes, it's not allowed. So, you can't pay yourself minimum wage and take $50,000 in dividends.

Again, don't mess with the IRS.

But also remember that you are allowed to take normal, reasonable, legal actions to reduce your taxes. Creating a corporation for your business is certainly a normal business activity.

I am not a tax professional. Don't do something just because I presented some ideas here. Find a qualified tax professional and go over the numbers. If you're in that $60,000-$127,000 range, you might save yourself some money.

At a minimum, don't dismiss the idea because it seems complicated or tax prep will become more expensive. A good tax pro will always save you money.

Define Yourself: S-Corp

When you decide to incorporate your business, you automatically form a Subchapter C Corporation or C-Corp. You must then file Federal form 2553 in order to "elect" to become a Subchapter S Corporation or S-Corp.

The primary difference between the two, for small entities, is that C-Corps pay taxes on the corporation's profits and stockholders also pay taxes on any dividend disbursements. With an S-Corp, the corporation does not pay taxes. It files a tax form to determine what the profit is, then that profit flows to the income calculation on the stockholders' personal income tax form.

In other words, with a C-Corp your income is taxed twice. There are other advantages for C-Corps that make sense for large entities. But for closely held companies (e.g., you, or you and a spouse, or you and one other person) the S-Corp is really the only option to consider.

The key benefit of an S-Corp is that you can avoid paying the Social Security or Self Employment Tax on a portion of your income. See the discussion of Sole Proprietorships, above. There are also several deductions available as an S-Corp that are not available as a Sole Proprietor.

Another major benefit of the S-Corp form of business is that the business is an entity unto itself. In fact, legally it is a person, which is bizarre. With a corporation, you personally have a relationship with the corporation and the corporation has a relationship with your clients. So you have this layer that protects your stuff from their stuff.

With a corporation, you have a certain level of liability protection. If someone sues the business, they can't normally get to your personal possessions. I say "normally" because you have to take certain actions to maintain this "corporate veil." If you treat the corporation like your personal ATM and don't treat it like a corporation, the courts are likely to say that you did not maintain it as a corporation. So that puts your possessions back out there for lawsuits.

S-Corps don't make sense when you're first starting out unless you got some great long-term, highly profitable contracts. Generally speaking, you need to maintain a certain level of profitability — and expect that to continue — before you incorporate.

Note also that corporate tax rates vary from state to state. There may be minimum taxes due, even in a year when the corporation loses money.

If you are considering a corporation, find a tax pro who deals with corporations. Agree to pay an hourly fee and sit down with your financial information and "run the numbers." Get a best-guess estimate of what you'd pay in taxes as an S-Corp versus a sole proprietor.

I discuss finding a tax pro later in the book, but let me take a minute to discuss that topic with regard to S-Corp elections. Some accountants and tax folks are heavily biased in favor of sole proprietorships. Sometimes this bias is based on the tax code realities of your state. But sometimes it's based on the fact that they just aren't up to speed on doing S-Corp taxes.

That's why I recommend finding someone who has several S-Corp clients. They will also have plenty of sole proprietor clients. As a result, they can give advantages and disadvantages of each side, with specific numbers to help make the decision.

Finding a great tax professional is extremely important. And we tend to become personally attached to these folks. But your primary concern needs to be your business. And it may be the case that you need to find "the next level" of skill and ability in a tax professional so that your business can move to the next level.

One final note on S-Corps. If you grow and are successful, you will eventually form an S-Corp. If you always stay just one person, you may choose not to incorporate. But if you hire people and start growing, there are too many advantages to not incorporate.

Define Yourself: LLC, Partnerships, etc.

Once you move away from Sole Proprietorship and S-Corp, there's whole alphabet soup of options available, including:

- General Partnership
- Limited Partnership
- Limited Liability Partnership
- Limited Liability Company
- and additional entities available in individual states

Partnerships are generally to be avoided. They are easy to set up, but they provide no protection of your personal stuff from your own business partner. So, if things go bad, they can go very bad. If you invest in real estate, revisit this issue for that enterprise, but not for your technical consulting business.

LLCs aren't bad. They act a lot like corporations, but allow profits to be allocated by means other than percent of business owned. They tend to be more expensive to create than a corporation, and state laws governing LLCs can be different from Federal laws.

Again, I sound like a broken record here, but you need to do your research and check with your tax and legal advisors.

Generally, you should have a solid business reason for choosing the form of your business. And for almost everyone, that choice will come down to being either a sole proprietor or an S-Corp. If someone suggests that you create some other entity, make sure you understand why.

It never hurts to get a second opinion.

Important Safety Tip: Signing Agreements

It is extremely important that you sign your agreements with your official title. For example, I sign agreements as "Karl W. Palachuk, President, Great Little Book Publishing Co. Inc." I sign as an officer of the corporation, not as an individual.

Again, the goal is to keep your personal stuff out of the business stuff.

Whether you have a corporation, a partnership, or whatever, you need to sign documents as an officer of that entity (president, managing partner, etc.). If you fail to do so — if you sign as an individual — you could create an opening that allows someone to "pierce the corporate veil" and go after your personal assets.

Concluding Comments

How you define your form of business will have little direct effect on how you write your Service Agreements. That's why we don't have sample language for each form of business.

There are several clauses, however that are more important for sole proprietors than corporations. We address these specifically in future chapters. These clauses address your independence in order to establish that you are not an employee of the client.

So, you've defined who you are legally. This decision will help you define client relationships. Remember to play your role at all times.

If you're a sole proprietor, act like a sole proprietor. If you're an S-Corp, act like an S-Corp. And so on.

Some clients will request a W-9 form, which is a statement relieving them of the responsibility of withholding taxes from your "pay." You should fill these out for anyone who asks. If you are a Sole Proprietor, you must give this information to each client. Consider filling out one form and photocopying it for each new client.

Bonus Business Tip: Invoice Labor Separately

When billing clients, send one invoice for labor and another for hardware and software. This allows the client to easily calculate the labor paid to you. That, in turn, makes it easier for the client to fill out 1099's at the end of the tax year.

This practice also allows the client to track hardware and software purchases more easily. Some companies depreciate these assets over five years, some choose to expense some or all in a single year. Keeping non-labor purchases separate makes the bookkeeper's job easier. In addition, this paperwork is useful for insurance, managing grants, and other activities your client may have.

You might even have a discussion with the client's bookkeeper to find out what works best. In addition to providing better service, you'll have a friend in the person who writes the checks!

"A verbal contract isn't worth the paper it's written on."
— Samuel Goldwyn

Chapter Two – Define Client Relationships

In the last chapter, you figured out who you are. Now you need to determine what your relationship is to your clients.

This chapter addresses the variety of possible relationships from "none" to very strong. Unfortunately, it is not possible to create a simple continuum with no contract on one end and extreme contract on the other.

Generally speaking, we define the following five types of relationships between the I.T. Consultant and the Client:

1. No Contract / Only Credit Agreement

2. Agreement for Hourly Work–Time and Materials

3. Agreement for Hourly Work–Blocks of Time

4. Agreement for Flat Rate Pricing

5. Agreement for Managed Services

While our primary focus is for technical support services, the same basic categories apply to software development or programming services.

And, finally, there are miscellaneous add-on services. These are things that don't fit nicely into the regular services. In particular, if you have a flat-rate pricing model for some services, there will be add-on pricing for specific "other" services.

We will look at each of these in turn. But first, some housekeeping.

Callout on Relationships: Business is Business, but . . .

Financial considerations aside, you should take time to consider the whole relationship with your clients. You probably want clients who are easy to work with, who pay their bills on time, and who are nice people.

You can make these criteria for your business. I decided many years ago to make working with nice people one of our criteria. We haven't had to get rid of too many clients over the years. And it has dramatically reduced our stress level and made our business an enjoyable place to work.

You should not lose sight of the fact that, at its core, your relationship with clients is financial in nature. Money matters, just as service matters.

But while you're making money, it's nice to be invited to picnics and birthday parties. It's nice to feel like you're a partner and a resource rather than simply an expense item.

So, don't forget that outside of your contracts you also need to work on the "softer" side of consulting.

Collect All Your Policies

If you are not currently using service agreements, then your client relationships are determined by a collection of "policies" you've put out there. These generally include:

- Text at the bottom of invoices
- Disclaimers at the bottom of quotes for service
- Notices sent by mail or included with your monthly newsletter
- Things you've posted on your web site
- Perhaps a 1-2 page agreement you had someone sign
- etc.

I hope you'll realize right away that most of these are not enforceable contracts. And, really, your clients are not likely to know 90% of what's in all these policies.

Perhaps you'll always want a line in a quote that warns that software licenses are not refundable. But for the most part, you need to collect all this stuff up and put it into a more organized and formal presentation.

When I get my bills (gas, electric, cable, etc.) I never read the inserts. I never read the back of the Visa bill. When I visit a gym and they make me sign a big disclaimer before using the equipment, I don't read it. I buy houses and never read the CC&Rs (covenants, codes, and restrictions).

We agree to software licenses we've never read. In fact, we agree to them by opening a package or accepting delivery. At some point, this is all just meaningless "legalese." Until somebody gets hurt.

At some level, we all want to cut through the legalese and just get to work. In fact, that's how most small businesses start. But then you have an incident that makes you realize that you need to define what you're going to do about that problem in the future.

As your business grows, you find yourself with a series of inconsistent relationships with your clients. They may all pay different rates based on when they first came to you. Some are billed once a month, others when the work is done.

In other words, your relationships with your clients consist of however you feel today (and how they feel today). At some point, this becomes unmanageable. Perhaps your business grows to the point where you have too many clients to memorize all the "deals." That's good. Or you hire an employee. That's a great motivator for standardizing your client relationships. Nothing helps you see how unmanageable this situation is better than trying to justify it to a new employee.

You start the process of standardizing your deals by writing down what you want your relationships to look like. Over time you've learned what works and what doesn't. Write down what does work.

Let's look at several pieces of the client-consultant relationship. I've got some notes on each of these, along with tips from my own experience.

Pre-Pay vs. In Arrears

Some people start their client relationship by getting a check up front. This is particularly important when you are just starting out. If you can get away with it, this is always good. But as you grow, and when you have repeat business, you will eventually begin billing people and having them pay you "in arrears," which simply means you bill first and then they pay.

Interestingly enough, the third stage of a relationship is often to go back to being paid in advance. Whether by pre-paying for a block of time or billing managed services in advance, many consultants go back to the pre-pay model.

Some clients simply insist on being billed with terms of "Due in 30 Days." Some of these people also then way until 30 days plus 29 days to pay. In other words, they are never a month past due, but seem always to be past due.

If you bill clients in arrears, you need to make sure you train them well. We don't put it in our contract, but we have a nice letter that explains very simply "We do not perform work for clients with past-due accounts." When everything's going smoothly they simply ignore this letter. But eventually something happens and they call us. At that point, we ask for the past due balance and a deposit on the current work.

You have to decide what's comfortable for you.

Getting the Agreement Signed

I've had discussions with consultants about the "tale end" of the sales cycle: They've done a walk-through, created a quote, revisited the client, and got a verbal agreement for a contract.

But the signed contract never shows up.

There can be many reasons for this. Perhaps you weren't really working with the decision maker. Perhaps the urgent situation isn't urgent any more.

Unfortunately, there's not much you can do at this point. You can't force someone to sign an agreement. But you can probe to determine the reasons for delay. In particular, if someone has agreed to your quote but then not taken action, you will eventually get an explanation.

It is important to probe for three very basic reasons. First, you need a yes or a no so you can move forward with your own planning. Second, if there are objections, you can answer them and proceed to getting the agreement signed. And Third, there may be no reason whatsoever for the delay.

Unless the client's need is urgent, this agreement may be important to you but completely unimportant to the client. After all, they just want to get the work done. Verbiage and "therefore blah blah blah" is not what's important to the client.

And you might not care why they say no. But you should never let a quote sit out there without a response. All quotes for service must result in a yes or no answer — even if it's, "Yes, but not for six months."

Many prospects will not return calls if they've given the job to someone else. But you deserve some answer, so keep calling.

What's in the Agreement and What's Outside the Agreement

Perhaps the most important decision you have to make before you start typing up an agreement is "what's included?"

For example, your managed service agreement might include all remote support but no onsite support. Or your regular hourly agreement might cover only time-and-materials projects but you can sell remote monitoring as a completely separate product.

Our company periodically goes through what we call a Product Audit. We usually do this late in the year. (Read: December. Sometimes between Christmas and New Year's.)

A Product Audit is very straightforward. First, make a complete list of every service you sell. A review of your billing software is helpful. In addition to hourly services, did you start doing something else this year? Many new service offerings come from client requests. Do you rent equipment? Do you have a flat-fee service for removing spyware? And so forth.

Second, write down all the products you've thought about offering. You can start out looking at the big picture and move into the details.

Yes, it's all service. But perhaps you'll list monthly maintenance, disaster recovery planning, domain registration assistance, and other items separately.

Once you have a complete list of the services you offer and the services you might offer, step back a bit and see where things seem to fit. As a general rule, you will find one broad category that represents most of your income. What is it? How do you define it? What's included and what's not?

That one product should be in a separate agreement from everything else. In fact, you may decide that you don't need official agreements for your other services. If all work is done either as part of your core product (for which you have an agreement) or for clients who have signed a Credit Agreement, then there's no point writing a bunch of service agreements for a bunch of services that don't amount to much of what you do.

In either case, basic hourly labor charges can be used for whatever's not in the agreement.

If your life is simple, like that, then there's really only one reason to start creating more contracts – when you offer a significant new service. For example, if you begin offering offsite backup solutions, you may want a nice document to address what's covered and how you'll proceed. Plus you'll want a nice paragraph making clear that you're not responsible for data loss, etc.

The maxim KISS – Keep It Simple, Stupid – applies here. If you have fifteen clients "on contract" and have eight different service agreements, you've made your life too complicated.

The more "managed services" or flat-fee pricing you have, the more important it is to spell out what's included and what's excluded. You don't want to have an agreement for all technical support that ends up costing your money.

No Contract / Only Credit Agreement

As a general rule, you will begin your relationship with a client in one of two ways. Either you'll start with a small job or you'll start with a big job. With big jobs, it's very easy to convince the client to sign an agreement of some kind.

With small jobs, it's harder to get a signature on a service agreement. But you should at least get a signature on a Credit Agreement. This basic agreement is very simple. It covers the issue about employee vs. contractor and it obligates the client to pay his bill.

The bottom line for any service agreement is "the bottom line" – money. Any agreement covers tax issues, payment terms, and other topics that all boil down to money.

On a small job, you need to have something in place. Your client may be reluctant to enter into a long-term contract or a full managed services agreement. But it's easy to get clients to sign a one-page Credit Agreement.

As presented in Chapter Four, the Credit Agreement presents a very simple relationship – the client agrees to pay his bills.

In many states, there are low limits on the amount of finance charge for unpaid bills. You can overcome these with the Credit Agreement. You can also charge a fee for late payments, a fee for returned checks, and restocking fees for returned merchandise.

We have never had any problems getting someone to sign the Credit Agreement. We simply say, "Before we get started, we need to have you sign this agreement. It basically says that we promise to bill you and you promise to pay us." The client gets a chuckle out of that and signs.

Make this document available for downloading in PDF format and make sure your technicians know how to get ahold of it.

I mentioned earlier that consultants rarely have problems with people not paying their bills, or disputing the work to be done. And I hope you never have any problems. But if you do, you'll appreciate having a document signed by the client that lays out your relationship in a clear and concise manner.

We go over the pieces of the simple Credit Agreement in Chapter 4.

Agreement for Hourly Work—Time and Materials or Blocks of Time

The basic agreement used by most consultants is for either "time and materials" or blocks of time. These are really the same thing with different payment methods. With a time and materials agreement, the client generally agrees to buy a certain minimum number of hours. In exchange for that promise, you give the client a discounted rate.

Unlike the basic Credit Agreement, the longer Service Agreements cover all the bases with regard to your relationship: tax matters, confidentiality, staffing, disputes, additional work—everything.

If you haven't been using service agreements, you will begin a two-pronged approach to signing new agreements: converting your existing clients and signing up new clients.

Getting an existing client to sign an agreement is usually accomplished by offering price stability. For example: "Our standard rate is now $120 per hour. If you sign a service agreement we can keep you at the $100 per hour rate. Otherwise, we'll have to raise the rate on January 1."

Notice that you need to raise your "standard" rate in order for this to be successful. See the callout on Raising Your Rates, below.

This approach is particularly effective on clients who haven't seen a price increase for some time. If a client hasn't experienced a price increase in three years, for example, they are probably prepared for one. At that point, they already have a relationship with you, they like you, and the agreement is really just a way to formalize your relationship.

You will also sell service agreements in association with a "project" of some sort. This is true for both existing clients and for prospects. It is easier with prospects.

If you are about to raise your rates, then using the strategy discussed above, combined with a project, will help sell existing clients on a contract.

With new prospects, the sale is actually easier. You start by casually mentioning your standard rate (the highest rate you charge to anyone) and then say something like,

> But if you sign a one-year service agreement we can go to $_____ per hour. That's about ten percent off. You'll only have to agree to 25 hours of support in the next twelve month period. And since this project is estimated at eighteen hours, you'd have no problem meeting that requirement.

Callout on Rates

I'm shocked when I talk to consultants and find out how little most of them charge. I've met people who charge $40-50 per hour! This is not for you.

I don't care where you live or how small you are, that's too low! I would consider $80/hr to be the lowest acceptable price anyone should ever charge for tech support.

Pick up the phone and make some calls. What do people in your city pay for:

- An Electrician
- A Plumber
- A Construction Contractor
- Labor at a Mom and Pop computer store
- Labor at a National Brand PC store
- Staples Tech Center
- Best Buy Geek Squad

If you've been in business for any amount of time—and you're any good—you should be in line with the highest priced company on that list.

If you've got some certifications under your belt and a good solid business under your feet, I recommend you start raising your rates. And don't stop until you're the most expensive consultant in town.

This is much easier than you might think.

First, you will have the complete support of your clients. Really. You lock them into the lower rate with a service agreement and make sure they know how much you charge for "full price." Your clients will brag about you to their contacts.

"This is Karl, our I.T. Consultant. He charges $150/hour!"

They don't mention that they only pay $120/hour. So this friend thinks your client is paying full price—and they see the client is obviously very happy. The result? The new prospect wants the $150/hour consultant.

The big fear in this area is that you'll lose business. If you have a sound business (and, again, you know what you're doing), you won't lose money. In fact, you'll get busier and you'll get better clients.

Try it. The next time you go out on a sales call, quote an extra $20 per hour. The most you have to lose is that one job. But chances are the prospect won't bat an eye. And you just made $20 times as many hours as they'll ever buy from you.

If you pay attention to what's going on during the sales visit, you'll see that the client is really buying you. If they like you and they like the way you present yourself, then the price is just a number to write down for future reference.

Many consultants I deal with are surprised when they raise their rates. Most consultants only raise their rates when they get too busy. They think that the increase will drive away business. In fact, the opposite is true. I know only one consultants who lost business because they raised their rates. I know dozens who got busier! And since this book was first publishing in 2006 I have received many emails from Consultants who raised their rates and only made more money.

Agreement for Flat Rate Pricing

It can be scary – and a little dangerous – to start out with flat-rate pricing. But, if you have any experience as a consultant, may want to try it. Having said that, let me back up a bit, define some terms, and then explain why I give this advice.

By flat-rate pricing I mean having an agreement that covers one or more services for a single monthly fee. For example, you might include all desktop operating system and software support at $X per month per machine. Alternatively, you might cover all remote support for a flat fee and charge time-and-materials for onsite support.

I don't recommend that you offer any flat-rate services until you're sure you're going to make money at it. Duh. If you're new to the business, you don't have a base from which you can estimate the cost of providing certain services.

If you offer something and get a few subscribers, you could get yourself in trouble fast. At a minimum, you could find that you make a little money, but you don't have the time and resources to go make more money.

Even experienced consultants need to be careful. Just because someone else is providing a service for a given fee doesn't mean you can. You better find out how they're doing it. Do they have tiered response levels? Do they have monitoring software, remote control software, and other tools?

Our first flat-rate offering was remote monitoring of servers. We used a home-grown collection of tools and Windows' Terminal Server to monitor client servers. We started out charging about $100 per month and then moved that to $150. Over time we were able to "upgrade" the monitoring we did as we adopted new tools.

We started out slow and made sure we never lost money. When we moved to an expensive, professional software package to provide total remote support, we paid for it with the already-existing monthly revenues from monitored servers. We made less as a result, but now had a product that allowed us to offer a wider variety of services.

So, the bottom line is Go Slow.

Test-drive your flat-rate services with one client. Then two and three, etc. Don't put yourself in a position to lose money.

Warning: Avoid "All You Can Eat"

One of the phrases you hear bandied about is AYCE or All You Can Eat. Please promise me that you will never be tempted down this road. Do not use the phrase and do not let anyone in your company use the phrase.

"All" the client can eat is all of your profit. I have heard many, many horror stories from people who sold AYCE contracts. Here's the problem: Whether they do it from ignorance or malice, there are clients who will take advantage of you.

There has to be a nice, clear line between what's included and what's not included. We use the terms "maintenance" and "Add/Move/Change." Again and again, I chant the mantra:

> **Managed Services covers the maintenance of the operating system and software. Adds/Move/Changes are extra.**

The primary driver of the managed services model is the concept that preventive maintenance is better for you and the client than break/fix or reactive support. You will prevent problems so you don't have to fix them.

Let's take it apart: Maintenance of the O.S. and software. That means maintenance, not making changes. So, installing new software is not included. Once new software is installed and working, then it will be covered as maintenance.

Use this phrase with clients and employees. Focus on preventive maintenance and avoid giving away labor that includes adds, moves, or changes.

Agreement for Managed Services

By managed services as outsourced technical support delivered under a service agreement that provides specified rates and guarantees the consultant a specific minimum income. As with any new terminology that emerges, this phrase wraps up a number of things that many of us have been doing for years.

Some people use the term managed services to mean flat-rate support or to mean remote support. But the term "outsourced I.T." makes more sense, in my opinion. What it really means is that a client has turned over his computer and network operations to a professional. And the "managed" part means that the focus is on maintenance and regularly scheduled work, not on break/fix.

And while you will probably never get away with an agreement that says your client cannot hire anybody else to work on the computers, the chances that he'll do so after signing an agreement with you are very slim. And you can add a clause that says all labor related to cleaning up changes made by someone other than you is fully billable at your regular rate.

Most people have come to understand managed services as some combination of services provided at a stable, predictable price. That doesn't necessarily mean flat-rate. But it does mean that the client doesn't have to worry about big surprises as the technology budget goes up and down.

Service Level Agreements

SLA's – Service Level Agreements – are just a bit different from the other service agreements. A service level agreement includes statements of performance. They are most commonly associated with managed services, but don't have to be.

SLAs commonly include the following guarantees:

- Response time
- Server uptime
- Amount of time that might be lost to spam, spyware, or viruses
- Hardware repair times

And what's the guarantee? Usually it's money. If you don't perform as advertised, you rebate a certain amount of money or a percentage of the bill. You're motivated to minimize downtime. The sales point is that the customer will prefer a company that guarantees their work over one that doesn't.

When I talk to evangelists about SLAs, most of the benefits they discuss are really the benefits of managed services. They focus on maintenance versus break/fix, and on flat-rate versus open ended labor agreements.

As you can tell, I'm lukewarm on SLAs. If you've been in business a while and you know how much it costs you to support a specific kind of client, you can design a managed services agreement that makes money. And if you've been doing managed services long enough to estimate downtime and rebates, then you can calculate an SLA that's profitable.

Having said that, here are some thoughts on SLAs.

The basic structure of an SLA involves promises and performance. As a rule, you will promise up-time and give rebates when you don't reach your targets. So, the first thing you have to do is figure out what you're willing to guarantee. Are you willing to guarantee desktop uptime? Virus security?

The second thing you need to determine is how much to rebate. When we toyed with the prospect of offering SLAs, we talked to several consultants who offer them. The consensus seemed to be that you should offer more than the client paid. In other words, a catastrophic problem for the client should also be painful for you.

In the samples here, we use a factor of three: One day of downtime equates to three days' worth of rebates. This is painful but not horrible. After all, a client would have to have 8-10 working days of downtime before you rebate the entire month. If they have that much downtime, you really aren't maintaining their systems.

But, you might argue, it's not your fault! The customer has old, junky equipment and won't buy new stuff. If that's the case, you need to consider who is eligible for SLAs. Perhaps you'll limit it to clients who have new equipment, or who agree to replace equipment on a three-year schedule.

Sample SLA Language

The following SLA language only covers servers. The covered servers must be listed.

Text:

Service Level Agreement (SLA) Attachment

1. Goals. The purpose of this SLA is to provide dependable, timely service and support for the technology needs of Client. Consultant agrees to respond in a timely manner and to resolve issues with the operating system and software on the server(s) listed below ("The Listed Servers").

2. Sole Provider. In order for this Service Level Agreement to be effective, Consultant must be the sole provider of technical services to Client. If technical support services are provided by any other party, including Client, the SLA portion of this Agreement will be invalidated.

3. Operational. Operational is defined as working and available to Client. A server is Operational when the Client can use it in the normal course of business and Client's business operations are not adversely affected

by the operational state of the server.

4. Failure. A failure occurs when any critical function of the server is not performing adequately to allow Client to continue normal operations. It is assumed that Client is losing money when the listed servers have one or more failures.

5. Exclusions. Failures under this SLA do not include hardware problems, including but not limited to problems with the motherboard, processor, disc drives, network cards, fans, and power supplies.

6. Force Majeure. Client acknowledges and agrees that Consultant shall not be responsible for any failures or delays in performing services under this SLA which are caused by actions or events outside of Consultant's control. Such actions include but are not limited to hardware failure, network interruptions, actions taken by Internet service providers or other third parties, Acts of God, acts of civil or military authority, fires, wars, riots, earthquakes, storms, typhoons, and floods.

7. Claims. In order to be compensated under this SLA, Client must report downtime and response time incidents in writing, by email, or by fax. Claims for SLA remedies must be made within 7 days of the failure.

8. Hours of Service. Consultant Hours of service are

8:00am to 5:00pm on business days, excluding national holidays. Consultant's help desk, web site, and off-hours voicemail serve as means of reporting service requests.

9. Response Times. All service requests for the listed servers will be acknowledged to Client within (hours) hours. Work on all service requests for the listed servers will begin within (hours) hours of the submission of the service request. Consultant will make periodic reports to Client regarding the status and progress on the service request.

10. Downtime Compensation. In the event that the Listed Servers experience unplanned Failures or are not Operational within the definitions of this Agreement, Consultant agrees to provide compensation to Client equivalent to three times the amount paid by Client for each day the server is affected. Compensation will be in the form of credit applied to Client's account. The minimum failure time will be one day and the minimum compensation time will therefore be three days.

Downtime percent is calculated as follows:

$$\frac{\text{Days or partial days of failure x 3}}{\text{Workdays in the month}}$$

Downtime Compensation is calculated as follows: Downtime percent times amount paid for server maintenance for the month.

11. Response Time Compensation. In the event that Consultant fails to provide adequate

response time, as defined in this Agreement, Consultant agrees to provide compensation to Client equivalent to three times the Ontime Failure Rate times the amount paid for server maintenance for the month. Compensation will be in the form of credit applied to Client's account.

Ontime Failure Rate is calculated as follows:

$$(1 - (\text{Number of ontime responses} \div \text{Total Service Requests in the Month})) \times 3$$

Ontime Failure Rate Compensation is calculated as follows: Ontime Failure Rate times amount paid for server maintenance for the month.

12. Liability. Client agrees that Consultant's entire liability, and Client's exclusive remedy, with respect to this SLA is limited to the amount Client paid to Consultant for Services during the previous three (3) full months. All other limitations of liability are identical to those agreed to in the Service Agreement.

13. Enrolled Servers. The following server or servers are covered by this SLA:
IN WITNESS WHEREOF, the parties hereto have executed this Agreement as of the date first written above.

Consultant:_____   Client: _____
By:_____   By: _____
Name: _____   Name: _____
Title: _____   Title: _____

Additional Options for SLAs

In addition to covering servers, you might consider covering almost any aspect of your service. This might include configuring firewalls and routers, fixing desktop PCs, monthly maintenance, etc.

Essentially, any job that you currently do could become a job with a guaranteed response time and a guaranteed uptime. You do not need to use the strict uptime definition used in the language above. If you calculate one hour of possible downtime per month for maintenance, for example, you can provide 99% uptime and not have to give a rebate.

I'm sure you can predict my next piece of advice: Just go slow. Design the system that works for you in your market. Don't just grab onto some thing and assume it will work.

Personally, I have never offered or signed a Service Level Agreement. I'm not in the business of giving money away. In fact, as you'll see, I have specific clauses in the contract that basically say we're going to try really hard but we're not responsible for anything that happens. More on this later.

Setting Prices for Managed Services

Everyone's got a different approach to setting prices for managed services. Here's what we did to begin offering a flat-rate component to our service agreement offering.

Whenever you make a big change like this, you need to "run the numbers" or estimate what the world will really look like if you go down this path. I say really because we all tend to fudge the numbers or nice-up our spreadsheets.

You need to be brutally honest here. Don't just round up the revenue a bit and flatten out the expenses. One of the goals here is to make sure you're profitable! Here are the basic variables you need to do that:

- Revenue from labor: Current and Projected
- Cost to deliver that labor
- Profit (do the math)

We assume that this is a labor-only offering: No hardware and no software. You could come up with something like that. You just have to run the numbers and make sure it works.

Callout: Delivery Costs

The cost to deliver is key. If you don't know what it costs to deliver one hour of labor, you need to take a detour and figure that out first.

Sometimes I talk to consultants who think they know this number but don't. For example, they do this calculation:

Bill 1 hour	$100
Pay tech 1 hour	($30)
Profit	$70

You need to look at two additional variables: operating costs and efficiency.

If this is an employee and not a contractor, you need to calculate taxes, unemployment insurance, benefits, cost of hiring, training, etc. Whether an employee or not, you need to consider all your overhead: rent, payroll service, office supplies, computer equipment, etc.

If you don't know these numbers, generate some annual reports from QuickBooks or whatever financial package you're using.

Round this number up. Go slowly. Be careful.

When you look at labor rates, calculate about 1.25 times the base rate to cover Social Security, unemployment taxes, etc. So, a rate of $25/hour becomes a true rate of $31/hour – if it's a 100% billable hour. At the same time, we know we're not 100% efficient. Sometimes we pay people to sit around. I'd give a lot of money to make that truth go away, but right now that's the truth.

We also have jobs where we give credit or simply don't bill for all the labor. We make sure the client knows when they're getting something for free, but that doesn't change the fact that we sometimes give away labor.

Again, calculate your worst-case scenario. Let's say a fulltime tech gets paid 2000 hours a year (we don't pay for vacations). Let's say that same tech bills about 30 hours a week or 1500 hours a year. That's an awesome technician!

But that means a worst-case scenario is around 20 hours a week or 1000 hours a year. That's a brand new technician (or one who's about to be fired).

Here's the calculation for base labor:

Base Pay	$25/hr
Taxes, etc.	$6/hr (about 25% of base)
Annual cost	$62,000
	($31 x 2000 hours NOT including overtime)
Efficiency	50%
Therefore bills	1000 hours
Bill rate	$100/hr
Revenue	$100,000

Here's the calculation for all non-labor costs:

Number of hours billed in last 12 months
Divided by total non-labor expenses
= General operating expense per hour of labor billed

For argument's sake, let's say this is $25.

And, finally, here's what it really costs to deliver one hour of labor:

Cost to deliver 1000 hours of labor is $62,000
PLUS general operating expenses of 1000 x $25 = $25,000
Grand total cost to bill 1000 hours of labor: $87,000
Total cost to bill one hour of labor: $87

Profit per hour billed is therefore $13 and not the $70 projected!

Your numbers could be dramatically different from this. The main point is that you need to know this number. It is very dangerous to proceed without it. If you don't have enough data to calculate this number, do the best you can. As you can see, you need to dig down and make sure you have very realistic expectations.

Note: Do not blindly take my numbers and start offering services. On one hand, you might not sell anything. On the other hand, you might sell too much and not make any money.

Okay, so you do some calculation and determine what it costs to deliver an hour of labor. Then you run some reports on labor sold in the last year and projected in the next year. For the following discussion, we'll assume you have access to your financial data and can run some reports by the client.

If this is useful, fine. If not, don't use it. You need to feel comfortable with your decisions here—they could be very expensive!

Go slowly. Plug in numbers for your market. Alright, enough warnings. Here's what we did.

In our case, we looked at our ideal clients. You might be tempted to pick your ten largest clients, but that's not necessarily the best choice. We looked at all of our clients and picked a handful that represented where we want our clients to be. In other words, we built our model around a group of clients that represented the ideal of what our next ten clients would look like.

So now we knew to whom we were selling. Next, we looked at historical numbers. We ran reports of labor sales per client for the last two years and exported that to Excel. Then we ran reports on the hours we paid for labor to support each client during the same period. We exported that to Excel.

Now we had two key figures: what we sold in labor and what it cost to provide that. See the callout above.

The great thing about this approach is that it includes all the labor. So, we could ask the question, "If we just came up with some big, huge number to cover all labor for a year, what would that number be?"

This is a good approach because your agreement should be as simple as possible. "Everything's included" is pretty simple.

At this point, we knew who our idea clients were, how much labor they used in a year, and how much it cost us to provide that. Next, we looked at every single labor sale over a two year period and tried to decide what would be included.

By default, everything was included, so we had to decide what not to include. Please refer to the spreadsheet on the next page as well as the Excel spreadsheet entitled Sample client labor sales.xlsx in the downloadable content.

First, we excluded projects. A project might be large or small. It might even take one technician fulltime in addition to regular maintenance. For example, we had a client hire us to set up a "ham radio" repeater that operates securely over the Internet. In addition to the equipment itself, we had to make changes to the firewall and coordinate with other "ham" users on the secure network. None of this falls under "maintenance" of servers or desktops. So, projects are not included.

Second, we looked at new hardware installs. Do we want to include a server migration? No. How about a new desktop? We decided to include the first three hours of labor on any new equipment. What about hardware failures? Out.

Pretty soon it became clear that we were covering **all maintenance**, but no new undertakings. And we were covering labor related to software and operating systems, but not labor related to hardware failures.

At this point we have narrowed down very nicely what it cost to provide all operating system and software maintenance for a year. We know how much revenue we generated in the past and what it cost to provide that.

See Diagram 2-1

Reviewing labor to determine what's included.

Finally, we created "model" clients in our Excel spreadsheet. For example, one server and twenty desktops, or two servers and twenty-five desktops. We created some inputs for what we would charge. Like this:

See Diagram 2-2

This is included in the Flat Fee Sample Clients.xlsx spreadsheet in the downloadable content.

As you can see, you plug in the proposed price per desktop, server, network, and "add-on server." We put the add-on server in because we figured that a server with Exchange (SQL, SBS, etc.) would take more labor to support than a basic file server.

Sample Client Labor Sales

Date	Num	Memo	Item	Hours Qty	Sales Price	Included Amount
01/31/2018	128331	Labor 1/14/2018: Gave Betty instructions on outlook F5and Web outlook. Connected Sid to th.	Labor Tech	2.50	$ 150.00	$ 375.00
01/31/2018	128331	Labor 1/18/2018: Worked on setting user names correct in Word and Excel. Only three users not d...	Labor Tech	6.50	$ 150.00	$ 975.00
01/31/2018	128331	Labor 1/24/2018: Installed RF Flow for John and Gwen. Spoke with Sheila re laptop and email	Labor Tech	0.50	$ 150.00	$ 75.00
01/31/2018	128331	Labor 1/25/2018: Set up Gwen. Betty and Stephanie to log onto Francie's PC and use the scam...	Labor Tech	2.25	$ 150.00	$ 337.50
01/31/2018	128331	Labor 1/26/2018: Help Andre Troubleshoot slow startup on laptop. MS Journal Reader was trying...	Labor Tech	3.75	$ 150.00	$ 562.50
01/31/2018	128331	Labor 1/26/2018: No charge	Labor Tech	4.50	$	$
01/31/2018	128331	Labor 1/27/2018: no charge	Labor Tech	0.50	$	$
01/31/2018	128331	Labor 1/27/2018: Set up printing drivers for the new Savin copier/printer/scanner. Remove all s...	Labor Tech	4.50	$ 150.00	$ 675.00
01/31/2018	128331	Labor 1/3/2018: Monthly maintenance. worked on issues in Baseline Security Analyzer report. Mon...	Labor Tech	3.25	$ 150.00	$ 487.50
01/31/2018	128332	Labor 1/3/2018: Server monthly maintenance. troubleshoot problems with the tape drive and resto...	Labor Tech	1.00	$ 150.00	$ 150.00
01/31/2018	128332	Labor 1/4/2018: RAID controller batter recondition. attempt to fix install for Windows. net fra...	Labor Tech	0.75	$ 150.00	$ 112.50
01/31/2018	128332	Labor 1/7/2018: Download and install critical updaters and vulnerabilities. Installed s-Helenet of...	Labor Tech	1.00	$ 150.00	$ 225.00
01/31/2018	128331	Remote Monitoring email and server - January. Server One	Labor Tech	1.00	$ 150.00	$ 150.00
01/31/2018	128332	Remote Monitoring email and server - January. Server Two	Labor Tech	1.00	$ 150.00	$ 150.00
02/14/2018	128652	Labor 2/4/2018: Complete setup of Ethel's laptop including office Pro. Windows and office upd...	Labor Tech	2.00	$ 150.00	$ 300.00
02/28/2018	128675	labor	Labor Tech	1.00	$ 300.00	$ 300.00
02/28/2018	128675	Labor 2/17/2018: Checked Backup Exec for problems. Francie could not put in a tape so we examine...	Labor Tech	0.50	$ 150.00	$ 75.00
02/28/2018	128676	Labor 2/17/2018: Checked on Backup Exec and tape drive. Discovered AppleTalk errors in event vi...	Labor Tech	0.50	$ 150.00	$ 75.00
02/26/2018	128676	Labor 2/2/2018: Monthly maintenance of server SM1 (web server).	Labor Tech	1.00	$ 150.00	$ 150.00
02/26/2018	128676	Labor 2/2/2018: Monthly maintenance of Server CLIENT2002 and Server2. Cleanup and tune up on...	Labor Tech	3.00	$ 150.00	$ 450.00
02/28/2018	128676	Labor 2/22/2018: Checked Backup Exec errors. Cleared some space on server. Reviewed updates n...	Labor Tech	1.00	$ 150.00	$ 150.00
02/28/2018	128676	Labor 2/23/2018: Checked Backup exec check. Picked up spare PC for disposal	Labor Tech	0.50	$ 150.00	$ 75.00

Diagram 2-1

Proposed Service Contract
Flat-Fee Support -- Software Support Only

Proposed Price Card

    Support

PCs / Laptops	$60	per month	equals	$ 720.00	per year	
Servers	$250	per month	equals	$3,000.00	per year	
Add-on Svr	$100	per month	equals	$1,200.00	per year	for SQL, Exchange, or SBS
Network	$250	per month	equals	$3,000.00	per year	Includes Internet Maintenance

The following sample prices are based on First Server with the $100 add-on and Network service.

PCs / Laptops	Servers				
	1	2	3	4	5
5	$10,800	$13,800	$16,800	$19,800	$22,800
6	$11,520	$14,520	$17,520	$20,520	$23,520
7	$12,240	$15,240	$18,240	$21,240	$24,240
8	$12,960	$15,960	$18,960	$21,960	$24,960
9	$13,680	$16,680	$19,680	$22,680	$25,680
10	$14,400	$17,400	$20,400	$23,400	$26,400
11	$15,120	$18,120	$21,120	$24,120	$27,120
12	$15,840	$18,840	$21,840	$24,840	$27,840
13	$16,560	$19,560	$22,560	$25,560	$28,560
14	$17,280	$20,280	$23,280	$26,280	$29,280
15	$18,000	$21,000	$24,000	$27,000	$30,000
16	$18,720	$21,720	$24,720	$27,720	$30,720
17	$19,440	$22,440	$25,440	$28,440	$31,440
18	$20,160	$23,160	$26,160	$29,160	$32,160
19	$20,880	$23,880	$26,880	$29,880	$32,880
20	$21,600	$24,600	$27,600	$30,600	$33,600

Diagram 2-1

The lower part of the sheet has the outputs. So, for example, a client with one SQL Server and eight desktops would pay $12,960 per year for maintenance of these systems.

Also on the enclosed Excel spreadsheet, you can see where we plugged in our current "ideal" clients and how the outputs compared. Then we played with various pricing schemes to see how the results compared to what we had been doing.

Of course, some clients paid more and some paid less than the projected total. We had to pick a number high enough to make sure they were all profitable. Because every client's different, we didn't find anyone who fit the ideal just perfectly. But we played with the numbers enough to know that we were in the right ballpark.

You should have a lawyer review every agreement or contract you sign.
www.smallbizthoughts.com

We refer to all of our service agreements as "Managed Services." This includes those charged by the hour and those that are flat-fee. The key to Managed Services is that we become the client's outsourced I.T. Department. That means we handle everything. Preventive maintenance is the most important process of making this work.

If all you want to do is break/fix and run around after the emergency of the day, then offering managed services is not for you.

What We Offer Our Clients

We have several packages available to our clients. The core offerings are:

1. Hourly Technical Support (signed credit agreement required). This is only for new clients. If they want ongoing support, they need to sign a support agreement.

2. Service Agreement for Hourly Support. Client receives reduced rate in exchange for agreement to purchase a specific number of hours per year.

3. Remote Server Monitoring. Available to clients with signed service agreements. Flat fee per month.

4. Remote backup of data. Fee per GB. Available to clients with signed service agreements. Flat fee per volume.

5. Fixed-Rate Maintenance of servers, desktops, laptops, and network.

6. HaaS – Hardware as a Service

7. Our Cloud 5-Pack

We have a price schedule for each of these. Includes all operating system and software maintenance and troubleshooting, whether remote or onsite (we decide). Includes all patches, fixes, and updates to all machines.

In my definition of managed services, all but the first item are managed services. Clients have turned over I.T. operations to us.

Concluding Comments

The bottom line on defining your client relationships is that you should do what is natural, comfortable, and profitable for you. As your business grows and evolves over time, you may decide to change the way you do business.

Do not feel obligated to offer managed services or redefine your business to fit the latest trend. But do use service agreements to formalize these relationships. There's just too much financial exposure to operating a business without any enforceable terms.

In the next section we discuss the "parts" that go into building service agreements. We also have some boilerplate examples of complete agreements of various flavors.

Section II
The Language of Service Agreements

Chapter 3
The Service Agreement Construction Kit:
Understand the Parts

Chapter 4
Boilerplate: A Sample Credit Agreement for
Clients Without a Service Agreement

Chapter 5
Boilerplate: A Sample Service Agreement for
Hourly Services

Chapter 6
Flat Fee or Managed Services Parts and
Sample Agreement

"The hardest thing in the world to understand is the income tax."
— Albert Einstein

Chapter Three – The Service Agreement Construction Kit: Understand the Parts

This chapter describes the language in your service agreement. We give a brief explanation of each paragraph. **This is not a legal discussion but a business-level discussion.**

The goal is to help you see:

> ➤ What this language does for you
> ➤ How this language affects your relationship with the client
> ➤ Options that are available for various sections

You'll see that the "parts" fall into the following categories.

Structural. These sections are the basic requirements for any agreement. These include the parties to the agreement and the execution (signatures).

External Requirements. External requirements do one of the following: 1) Fulfill a client requirement (e.g., drug free workplace); 2) Provide clauses to make sure your agreement is structured the way you want it for tax purposes (see the section on Keeping Uncle Happy); or 3) Provide compliance with other state or federal laws.

Managing the relationship. These sections cover communication between the client and consultant, including dispute resolution.

The Deal. Here's the "meat" of your service agreement. These sections define what you'll do, what they'll pay, who owns what, etc. In any contract, each side gives something and gets something.

It doesn't really matter which category any given paragraph falls into. I just want to give you a sense of the kinds of sections you'll see below. I've presented a proposed agreement, organized in an order that makes sense to me. You or your lawyer may prefer another logical sequence.

Fitting into Your Existing Agreements

If you've been in business very long, you probably have some agreements you've pieced together over the years. It might be a full "contract," or a series of agreements about credit terms, data integrity, etc.

To piece these parts together, you can start by making a new document with your own parts and the parts from this book. Just be aware that Frankenstein's monster "worked" but caused a great deal of trouble for its creator! After you mix and match various paragraphs from various agreements, you need to take your Franken-Contract to a lawyer who can re-arrange it, verify consistency, make sure no sections contradict one another, and make sure the final agreement will hold up to state law.

Note: In many jurisdictions, the benefit of the doubt is given to the contract party that did not write the contract. That means, if there are any problems or disputes, you'll be at a disadvantage because you drew up the agreement. That's one more reason to have a lawyer give it a once-over before you present your agreement to a client.

Keeping Uncle Happy – IRS Requirements that Affect You

Rule Number One: The U.S. Government will get its share! Rule

Number Two: If you don't define your business arrangements carefully, the government will get more than its share of your business.

Having a properly-drafted service agreement is a lot like having a will. If you die without one, the government has a standard set of procedures to follow. They result in long delays and very likely do not result in you having your wishes respected.

Similarly, if you operate a business without service agreements, everything you do is governed by the "default" common law, rules, regulations, laws, and procedures. At tax time, that can leave you at a real disadvantage.

Within your service agreement, you have the opportunity to spell out "answers" to the IRS's questions about how your business is structured. This is particularly important for sole proprietors. But it is equally important to your clients.

The major area of concern is whether you (the contractor) are really an employee of your clients. If the IRS determines that you are an employee to one client, they can examine your relationship with other clients. They can also hold your client liable for tax withholding, as well as back taxes and penalties.

So the bottom line is clear. Both you and your clients benefit from defining your relationship with a service agreement.

Some of the rules that the IRS uses to define employees are found in publications 15-A, 15-B, and 1976. You should also look at Form SS-8, "Determination of Worker Status for Purposes of Federal Employment Taxes and Income Tax Withholding."

Many of these forms, and related IRS materials, are included on the CD that accompanies this book. These things change all the time, so make sure that you check the IRS web site for the latest info.

Callout: Focus on the IRS

Make a little extra time to visit the IRS web site. Spend time getting an idea of what the government expects of you.

Go to www.IRS.gov.

More specifically, go to the "business" area at www.IRS.gov/businesses.

Links change all the time, but look for the following:

- Employment Taxes for Businesses
- Forms and Publications
- Filing Late / Paying Late
- Tax Code, Regulations and Official Guidance
- Small Business / Self-Employed
- Operating a Business

And here's a bonus section on the IRS web site: **Small Business Products and Online Ordering**. There you'll find some great links to free products, including:

- Tax Calendar for Small Businesses
- Small Business / Self-Employed Electronic Marketing Card
- Virtual Small Business Workshop CD
- Small Business Resource Guide CD

Pay particular attention to the section on "Independent Contractors vs. Employees." This is a major ongoing concern for the IRS.

Don't mess with the IRS!

Let's assume before anything else that you are willing to pay your taxes. You may not be happy about it, but you're willing to do it.

That's the first step in keeping Uncle Sam happy.

I'll admit that I'm risk-averse when it comes to paying taxes. I've seen people go out of business, go bankrupt, and still have a large IRS bill to pay. You can't escape paying taxes, and you really shouldn't even delay. If you have trouble keeping up with the quarterly tax payments, that's a sign that you need to make some changes fast.

Having said all that, let me give you the other side of the story. My parents owned a tax preparation business and I'm a firm believer in the federal statute that states that you have the right to use every legal means to minimize your federal income taxes. In fact, I argue, you owe it to yourself and your family to minimize your taxes.

Very often, a simple decision can affect thousands of dollars in taxes. The most obvious example is whether to operate as a sole proprietor or an S-Corp. There are a handful of tax-related issues that you need to address in your contracts. These include:

- Your status as a contractor with regard to your clients
- Your status as an employer with regard to your employees
- Your office space (home office, lease office space, etc.)
- Reimbursable expenses (e.g., mileage, "tools")

As a sole proprietor, the most important consideration is your status as a contractor and not an employee of your clients. This is an area of law that developed significantly since 2000, due in large part to the use of contractors in our industry.

Many dot.com companies used contractors in order to avoid paying benefits, taxes, workers comp, and other employee-related expenses. In some cases the IRS challenged these contractors' status. In other cases, workers sued companies that treated them as employees but paid them as contractors.

At the same time, the IRS went through phases of cracking down on home office deductions. These two are actually related. One of the criteria for determining employment status is whether you have a workspace dedicated to you and one of the criteria for taking a home office deduction is whether you have another place to do your work.

The next section is entitled "The Parts" as it contains sample agreement language for various pieces of a service agreement. Each part is accompanied by a description or discussion.

Note: I'm not an accountant or Enrolled Agent. Just like I'm not a lawyer. This is all just information. Nothing here is to be taken as tax advice. You're responsible for your own actions. If you use these paragraphs to create a service agreement, you should have it checked over by a lawyer and a tax professional. For most of you, Cousin Larry is neither, so his advice doesn't count.

The Parts

Here we discuss the specific sections of your Service Agreement. In the next few chapters we'll look at samples of entire documents. Each agreement "part" is presented in fixed-width font, followed by a discussion of the paragraph.

A simple "x" is used rather than a section number. When you piece together your agreement, you will obviously replace these with appropriate numbers.

The Text:

```
This  Agreement  is  made  as  of  ____[ date ]___.
between _____ ("Consultant")
And _____ ("Client").
.  .  .
```

x. Notices.

(i) Notices to Consultant should be sent to:
    [Address]

_____

_____

_____

(ii) Notices to Client should be sent to:
     [Address]

_____

_____

_____

. . .

IN WITNESS WHEREOF, the Parties hereto have signed this Agreement and agree that it shall be binding upon the parties and their respective heirs, successors, and assigns.

Date Signed: _____

[sign]

_____

Consultant
(You, your title)
(Your company)

[sign]

_____

Client
(Name, Title)
(Company)

Discussion:

These are the basic "housekeeping" sections. Pretty straightforward. The part before the ellipses (. . .) goes at the top. The part after the ellipses goes at the bottom of the agreement.

See the discussion in Chapter One about signing agreements, especially if you represent a corporation.

```
The Text:

x. Term of Agreement. This Agreement for
Consulting Services is for the Period _____
to _____. This agreement may be extended
by execution of an extension by both parties.
Either party may cancel this agreement upon
thirty (30) days written notice.
```

Or:

```
x. Term of Agreement. This agreement shall
commence on the date set forth above and will
continue until either party hereto provides
the other with written notice upon thirty
calendar days' notice of termination.
```

Discussion:

The term is pretty basic. The agreement has a beginning and an ending date. Alternatively, you can have an agreement that continues forever.

You are probably predisposed to one of these. I started out with date certain agreements. I didn't really have a good reason. In 1995, most people were not signing any contracts for computer consulting services, so I thought it was best to lay out the deal very precisely. I could summarize the deal in one sentence:

In exchange for a better hourly rate, you are agreeing to purchase at least _____ hours over the next twelve months.

That led me to defining the twelve months. I had just come from a rather corporate world and was used to negotiating "deliverables" and dates.

Some work, especially project work, is best performed with a series of deliverables and due dates. After more than 25 years in the service agreement business, we now prefer the more open-ended agreement. We ask clients to agree to a certain number of hours per year, but we don't really enforce that.

In the final chapter we discuss how you keep track of all the service agreements, rates, hours, minimums, etc. Generally speaking, keeping track of all that becomes easier if you use open-ended agreements.

The Text:

```
x. Early Termination. Client recognizes that
the preferred rate defined in this Service
Agreement is contingent upon the purchase of
a specified number of labor hours. In the event
that Client terminates this Agreement, for
any reason other than failure to perform the
services outlined herein, before the Client
has purchased the number of hours agreed upon,
Client agrees to pay an early termination fee
of $ _____.
```

Discussion:

If you have a flat-rate agreement, you might consider the early termination to be equivalent to one or two month's payments.

The goal here is to make sure a client does not take advantage of you. If someone takes your 10% off and then only buys five hours, they don't really deserve the discount! Again, we've never really had any problems with this. This is a "good fences make good neighbors" clause. We both agree on the boundaries and it's not an issue after that.

Another option is to require a minimum number of months per device. So, for example, I require a four-month minimum. There are hard costs to supplying managed services, installing RMM agents, etc. The four-month minimum basically guarantees that I can't lose money. Here's the sample text for that:

```
Client recognizes that the preferred rate
defined in this Service Agreement is contingent
upon the purchase of a specified number of labor
hours. In the event that Client terminates
this Agreement, for any reason other than
failure to perform the services outlined
herein, before the end of the fourth month,
client agrees to pay the amount that would be
due for the first four months of service.
```

The Text:

```
x.  Complete  Agreement.  This  Agreement
constitutes  the  entire  Agreement  between
Consultant and Client. No other representations,
whether written or oral, shall bind the parties
with regard to the matters agreed to in this
Agreement. Consultant and Client acknowledge
that they are executing this Agreement based
solely on the written text of this Agreement.
```

Discussion:

This section makes clear that this agreement only covers the specific items mentioned in the agreement. You may have additional agreements for remote monitoring, web hosting, managed services, etc. If you have specific products that have separate agreements, make sure that you don't mention them in this agreement.

The basic approach is to put each activity into "compartments." You can then decide to have an agreement for each, or to have no formal agreement for some. Note also that any promises you make in advertisements are not part of this agreement. Of course that won't stop your clients from holding you accountable for promises you make!

The Text:

```
x.  In  the  event  of  a  conflict  between  the
text  of  this  Agreement  and  the  text  of  any
Attachments  to  this  Agreement,  the  provisions
of  the  attachments  will  prevail.
```

Discussion:

You could, of course, reverse this and make the main agreement override the attachments. I like to have the attachments prevail because that makes it easier to put together a new agreement. I can print off the standard agreement with no changes and then put together an attachment with the specifics for the current client.

If you actually know you have a "conflict" between these two parts, it is best to resolve it and then have your attorney give it a look. Additionally, if you find yourself putting some important thing in an attachment to every agreement, it's time to integrate that information into the main boilerplate text.

We use "attachments" to execute Managed Service agreements. Since the attachment overrides the basic Agreement, we know there are no problems with this arrangement.

The Text:

```
x. Jurisdiction. The enforcement of this
agreement shall be governed by the laws of
the State of _____.
```

Discussion:

This is pretty standard language to determine which state or commonwealth will govern the agreement. If you don't know the difference between a state and a commonwealth, you live in a state.

The Text:

```
x. Enforcement. If any piece of this agreement
is held to be unenforceable or not fully
enforceable, then the parties agree that
the remainder of this Agreement will remain
enforceable and unaffected. It is the intent
of the parties that this Agreement, and each
of its parts, will remain in effect to the
extent that they not affected by legal or
judicial findings.
```

Discussion:

Since you (and presumably your client) want this agreement to hold up under legal scrutiny, this language simply states that you intend for the agreement to be as enforceable as possible. The very best language for your state or commonwealth may vary.

Note that, if you use the arbitration provision below, the likelihood of any legal disputes is much diminished.

The Text:

```
x. Dispute Resolution. Should a dispute arise
with regard to the enforcement or application
of any provisions of this Agreement, the parties
agree to binding arbitration by a mediator
associated with the American Arbitration
Association.
The mediator's arbitration decision will be
enforceable according to the laws of the State
of _____.
```

Discussion:

As presented here, this dispute process just says that you will use The American Arbitration Association dispute resolution process. This process can actually be started (and sometimes completed) online.

If you wish to have disputes resolved by a mediator who is a member of the AAA, you can find one at https://www.adr.org/.

See the callout on arbitration and The American Arbitration Association.

Callout: Arbitration and
The American Arbitration Association
http://www.adr.org/

The American Arbitration Association (AAA) exists to help people save some money on expensive litigation. If one road to settling your differences is to hire a lawyer, then the alternative is ADR — Alternative Dispute Resolution.

Definition (from the AAA web site): "Arbitration is the submission of a dispute to one or more impartial persons for a final and binding decision, known as an 'award.' Awards are made in writing and are generally final and binding on the parties in the case."

What that means: If you and a client have a disagreement, the binding arbitration clause means that you both agree that you will abide by the decision of an arbitrator.

The American Arbitration Association is an 80-year-old organization that helps people settle their differences without the "open-ended" nature and cost of lawsuits.

To find a mediator, go to www.mediate.com and put in your state.

As with any relationship, it is best to do whatever you can to settle problems before going to arbitration or going to court. Once you get that far, your business has suffered more than it needs to.

Having said that, arbitration is usually the fastest, least expensive, and least painful way to settle disputes.

The Text:

x. Services Provided. Consultant will provide the following Services under this agreement:

Item y. Consultant agrees to provide consulting services related to computer hardware, software, network configuration, operating systems and networks, database development, programming, and other similar work. Such work will be done on behalf of Client and will be performed at a location or locations to be determined by Consultant. For example, services might include installing and configuring Windows operating systems, Internet servers, and associated software.

Discussion:

This is simply a broad listing of services provided. Unless there's a reason to do otherwise, this section should be as broad as possible. This is true for several reasons. First, you want everything you do to be included so you don't have to amend the agreement in order to perform some task.

Second, the technology business is a fast-moving industry that never stops changing. You don't want to be stuck in the past, unable to work on newer technologies. The following section is directly related to this:

The Text:

```
Item y. Regular Consulting Hours

Regular Consulting Hours shall consist of
any time Consultant works for Client during
regular business hours. Regular business
hours are defined as 8:00 AM to 5:00 PM Monday
through Friday excluding national holidays.

Item y. Extended Consulting Hours

Extended Consulting Hours shall consist of
any hours worked by Consultant outside of the
period defined as "Regular" Consulting Hours.
This includes weekends, the period 5:00 PM to
8:00 AM during weekdays, and all holidays.
```

Discussion:

We use the terminology Regular and Extended Consulting Hours. You may prefer some other phrases. Basically, you need to spell out that there are normal hourly charges and there are after-hours charges.

You may wish to list specific days you are not working. For example, if you're closed the day after Thanksgiving, you should list that.

Of course you will also need a section that specifically states that Extended Hours cost more than Regular Hours. See "Costs" below.

The Text:

```
Item y. Programming Hours

Programming Hours shall consist of any work
by Consultant using a programming language
such as (but not limited to) PHP, HTML, Visual
Basic, VBScript, C/C++, or Java.
```

Discussion:

If you provide programming or other special services not covered in the Nature of Services section above, list them here. You may charge the same rate or a different rate for these services. Listing them separately gives you the flexibility to charge a separate rate.

The Text:

```
x. Additional Work. Upon request from Client
to perform services, whether specifically
listed in this agreement or not, Consultant is
authorized to perform such services and Client
agrees to pay Consultant for such services at
the rates specified in this Agreement.
```

Discussion:

This section covers one of the most basic and most important elements of your contract. When the client asks you to do something, you expect to get paid for that. This seems obvious, but I'll give three examples of why you need to actually state it.

First, consider the case of the client who calls you out for a minor task. Let's say you get a call because a monitor seems a little too dim. You are in the middle of something but the client says he needs you out right now. You go, you adjust the monitor, and four minutes later you're finished. But you have a one-hour minimum.

Until they are "trained," some clients will be upset that you are charging so much for such a minor task. Of course, we recommend that you tell the client "You've got 56 minutes left on the clock. Is there anything else I can do while I'm here?"

A second example would be a request for a policy. For example, the client decides to have everyone clean up their email folders and asks you to draft a policy and procedure that can be distributed to the staff. You create the document and email it to the client. When a bill shows up for this work, the client says "Hey, you didn't do any work here. All you did is send an email."

Believe me, this kind of thing happens. You have to explain that they made a request and that, as a result of that request, you drafted a document. This takes time. Time is what you sell.

A quick third example is the cancelled project. A client asks you to begin work on a project and then decides not to proceed with it. The client might argue that she didn't get any value from your labor because the project never proceeded far enough. But the bottom line is that she asked you to do something and you did it.

Virtually all conflicts in this area can be avoided by the proper setting of expectations. Simply tell the client, "I'll come out, but I have to charge my one-hour minimum." Or agree to the policy-writing project and say something like "I'm sure the charge won't be more than an hour."

I know this seems like a long discussion for such a short paragraph, but this really is a key piece of your agreement. The bottom line is that you want the client to rely on you for more and more things. This section just makes sure you'll get paid.

I find it very interesting that some clients put no value on the "consulting" piece of computer consulting. They have no problem paying for labor they can see. But if you're just explaining something or giving advice, that's where they say "We were just talking." Just remember your mantra – time is what you sell.

Perhaps the sweetest words you'll ever hear in an initial conversation with a client are "I know your time's worth money." Give that prospect five free hours on the spot! Do whatever it takes to sign him up. Once he's on board, this is a client who will pay you to talk about how to improve his business. They should all be like that!

The Text:

```
x. Cost of Services.

All General Consulting Hours, as defined in
this Agreement will be provided at a rate of
_____ dollars ($ _____ ) per hour.

All Extended Consulting Hours, as defined in
this Agreement will be provided at a rate of
_____ dollars ($ _____ ) per hour.

All Programming Hours, as defined in this
Agreement will be provided at a rate of _____
dollars ($ _____ ) per hour.
```

You should have a lawyer review every agreement or contract you sign.
www.smallbizthoughts.com

```
These rates are subject to change, subject to
a thirty (30) day written notice.
```

Discussion:

Earlier we defined the scope of services. Here we put a price on it. In just a little bit we'll put some restrictions on this. The next chapter has an example of how this is filled out.

You may not have an evening rate (extended rate). Even if you charge the same rate for both, I would list regular and extended hours. Everyone knows that after-hours support costs more. If you don't charge more now, having this paragraph will give you an easy way to increase that later.

The Text:

```
x. Confidential Information. Each Party hereto
("Such Party") shall hold in trust for the
other Party ("Such Other Party"), and shall
not disclose to any non-party to the Agreement,
any confidential information of Such Other
Party. Confidential information is information
which relates to Such Other Party's research,
development, trade secrets, or business
affairs, but does not include information which
is generally known or easily ascertainable
by nonparties of ordinary skill in computer
design, programming, networking, information
technology, or the specific business interests
of either Party.
```

Discussion:

I'm a real believer that you can use your "standard" non-disclosure section as a selling point. It's just built in. When clients mention something about trade secrets or HIPAA compliance, you can casually mention that "Of course we're under a confidentiality agreement."

You also need to have your employees sign a confidentiality agreement before they touch a customer PC. Then you can assure your clients that this section applies to the entire consulting firm and not just to you.

Sample Agreement of Confidentiality for Employees

I, _____ agree to be bound by the following:

While performing any work as an employee of (Your Company), I acknowledge that I may come into contact with or possession of information belonging to (Your Company) or end-clients of (Your Company).

I acknowledge that all such information, whether disclosed by demonstration, writing, within programs, within databases, or otherwise is the property of its creator and is deemed confidential and proprietary information.

I will use such confidential and proprietary information solely for purposes of fulfilling tasks necessary for my work as an employee of (Your Company).

I will keep such confidential and proprietary information in strict confidence and prevent all unauthorized access to, use of, and disclosure of confidential and proprietary information.

You should have a lawyer review every agreement or contract you sign.
www.smallbizthoughts.com

All obligations under this Agreement of
Confidentiality shall remain in full force
during the term of my employment and
for a period of two (2) years after my
employment is terminated for any reason.
Remuneration for this agreement shall
consist of my regular pay as an employee
of (Your Company).

Signature _____

Date _____

Employee's Name (typed or printed)

Discussion:

Please note that you have made this agreement apply to both your
business and your clients' businesses.

In most cases, you need to exchange something of value in order to
have an agreement be binding. That's why you state that payroll is
being exchanged for confidentiality.

Photocopy these agreements and be able to produce them for
clients. You may never be asked for them. But when you find that
one who needs it, they'll want it to show up sooner rather than later.

The Text:

x. Staff. Consultant is an independent
Contractor and Consultant is not employed by
Client. Consultant agrees to provide services
as a contractor under this Agreement.
Consultant will determine which services are
to be performed, and the order and method

of providing those services in response to requests from Client.

Consultant provides services to a variety of clients and is not required to provide full time availability for services to client. Consultant agrees to pay all taxes related to the work performed under this Agreement. Client is not providing any insurance coverage on the Consultant or Consultant's staff.

Discussion:

If this sounds like a laundry list of criteria the IRS has been known to use when determining the status of an independent contractor, that's because it is. See the discussion of "Keeping Uncle Happy" earlier in this chapter.

Remember, this is a bit of a moving target.

In some ways, this section may be the most important one in the agreement. After all, it takes one job to determine whether your client is going to pay his bills. So agreeing on rates is important, but it would get settled anyway. Having the IRS come down on your client and ask them to pay withholding taxes as well as back taxes – that's important. Make sure your client knows this.

The Text:

x. Non-Solicitation of Employees. During the term of this agreement and for a period of twelve (12) months after the termination of this agreement for any reason, Client agrees not to recruit or employ any employee of Consultant without the prior written consent of the President or Chief Executive Officer of

```
Consultant. Consultant hereby agrees that it
will not recruit or employ any of Client's
employees without written consent from Client.
```

Discussion:

This is another section that is rarely needed, but you'll be grateful you have it when the issue arises. On one occasion, we had to deal with someone talking to our techs about work aside from my company. That is, rather than offering a job, they just wanted some work "on the side." A quick phone call takes care of that.

In general, you need to make sure your techs keep an arms-length relationship with your clients.

Some people add a monetary penalty to this, such as three times the employee's monthly salary. Unless you work in a climate where cross-hiring is a real issue, don't make a big deal of it.

The Text:

```
x. Drug Free Workplace Certification. Consultant
agrees to comply with the provisions of
(State Law or regulatory section) regarding
maintenance of a Drug-Free Workplace.
Consultant agrees to notify its employees
that they are prohibited from engaging in
the unlawful manufacture, distribution,
dispensation, possession or use of controlled
substances.
```

You'll want to tweak this a little to make sure it works in your jurisdiction. Your attorney or client may want to list the specific provisions of the law. You'll also need to draft a drug-free workplace agreement for you employees to sign. Here's what we use:

Drug Free Workplace Policy

(Your Company) maintains a drug-free workplace in accordance with the provisions of the Federal Drug-Free Workplace Act of 1988 and the (Your State's) Drug-Free Workplace Act of (year) (Your Company) strictly prohibits the unlawful manufacture, distribution, dispensation, possession, or use of any controlled substance in (Your Company's) workplace, including client job sites. Employees who violate this policy will be subject to serious disciplinary action, up to and including immediate termination of employment.

Drug Free Awareness Program

(Your Company) has also established a drug-free awareness program pursuant to the Federal and (Your State's) Drug-Free Workplace acts. It is important that employees understand the dangers of drug abuse in the workplace and (Your Company's) policy of maintaining a drug-free workplace. Employees who believe they have a drug or substance abuse problem should seek appropriate counseling or participate in drug abuse assistance or rehabilitation program approved for such purposes by a Federal, State, or local health, law enforcement, or other appropriate agency.

Requirement to Report Conviction for Violation Occurring in Workplace

Any employee who is convicted for violating any criminal drug statute based on conduct that occurred in (Your Company's) workplace, including client job sites, must report such conviction to (Your Company) no later than 5 days after such conviction. Such employee will be subject to disciplinary action, up to and including termination. Alternatively, or in addition to disciplinary action and in its sole discretion, (Your Company) may require such employee to satisfactorily participate in a drug abuse assistance or rehabilitation program approved for such purposes by a Federal, State, or local health, law enforcement, or other appropriate agency.

Every employee must acknowledge receipt of this policy and agree, as a condition of employment, to abide by the policy's terms.

I, _____, hereby certify that I have received and read (Your Company's) Drug-Free Workplace Policy and agree to its terms. I understand that any violation of the policy may result in serious disciplinary action, up to and including immediate termination of my employment.

Date:_____

Employee Signature:_____

Discussion:

With the massive changes in marijuana laws in the last several years, you might want to avoid this entire discussion altogether! Having said that, you may have a client that requires some sort of "drug free workplace" policy.

The laws are evolving quickly in this area. It may or may not be legal to limit the legal consumption of a legal drug. But, of course, you don't want your employees showing up high any more than they would show up drunk.

So the advice here is: Ask an attorney! In particular, ask an attorney that specializes in employment law.

Of course, your clients will be well aware of the evolving laws as well. So, sit down with them and discuss what they need. If they have proposed contract language, take that to your employment attorney and see how you can fit it in your contract.

The Text:

LIMITED WARRANTY

x. Liability. Consultant warrants to Client that the services and labor performed as part of this Agreement will be of manner and quality associated with a professional technical consultant.

Consultant offers no guarantees or warranties as to system availability and functionality during any phase of its support services and makes no guarantees or warranties regarding the ability to resolve computer-related problems, to recover data, to avoid losing data, or to prevent loss of income.

Consultant makes no other warranties that products delivered under this Agreement are fit for merchantability. In no event shall Consultant be liable to the client for any loss of profit, loss of business, consequential, or any other damages as a result of work performed under this Agreement. In the event that this limitation of damages is held unenforceable then the parties agree that all liability to Client shall be limited to ($ _____ .00).

KP Note: I recommend the number here be $10,000. See below.

> **Note: DO NOT copy and use this text without reading it.**

Discussion:
This is literally the "cover your butt" section.

First, this text basically says that you're not responsible for anything you do. You don't promise to fix anything. You don't promise that the client's computers will work. If you write code or help create some product, you don't promise that any of it is worth selling to someone. Period.

Second, if everything goes horribly wrong, you limit your exposure to the amount of money listed here. I use the number $10,000 because it's large enough that a judge or mediator will take it seriously. It's also small enough to not bankrupt me.

If ten thousand dollars is going to bankrupt you, then you need to lower that number. Do no spit out some agreement without giving serious consideration to this number.

Even if you offer a service level agreement and give rebates based on downtime, I think you still need this section and you need to limit your liability as much as possible.

As for the dollar amount, consider two points. First, consider what the client will say if he actually reads the agreement. If you limit your exposure to one hundred dollars (which I did for many years), you might have to explain to clients that you really are good, even though it looks like you don't stand behind your work. Actions speak louder than words, and you can always get a long list of client references.

But, second, you should make this number large if you can. This looks much better to the client. Some consultants tie this amount to services rendered. So, for example, you might say:

```
. . . all liability to Client shall be limited
to the total dollar amount of services paid
by Client to Consultant in the previous three
full months.
```

This allows you to look like you take the liability issue seriously, and the client sees an immediate fairness here. If he bought $10,000 worth of services, then the limit is $10,000; if he bought $1,000, then it's $1,000.

As I mentioned above, we moved this number to a flat $10,000 because it is large enough to be taken seriously and we hope it boosts the enforceability of the entire agreement.

I don't want to put too sharp of a point on it, but please take this section very seriously and discuss this number with your attorney. Also, please read the discussion of liability insurance in Chapter Seven.

The Text:

```
x. Terms.

1) Client agrees to purchase a minimum of
        _____ hours of Services during the Term of
this Agreement.
```

2) Client agrees to pre-pay _____ hours of labor upon execution of this agreement.

3) All invoices to Client shall be due within _____ days.

4) Any unpaid sums over _____ days old that are not in dispute shall bear interest at the rate of _____ percent per month. Costs of collection including reasonable attorney's fees shall be borne by the Client.

5) There is a _____ minimum charge for all onsite visits for services not covered by a separate managed services agreement. There is a_____ minimum charge for remote support services not covered by a separate managed services agreement.

Discussion:
Okay. Here's that collection of "other stuff" we discussed in Chapter Two. I left the 1, 2, 3, etc. in there for discussion purposes.

Note: The normal format is to spell out numbers and then print the numerals in parentheses, thusly:

one hundred dollars ($100.00)

1. How many hours are you requiring the client to buy? You might give one price for a commitment of 250 hours, one for 100 hours, etc. We generally set the threshold low and try to get them on board. Clients who intend to buy 250 hours are the ones most likely to be open to a managed service offering, with flat rate.

2. Pre-payment should be an option in your agreement even if you don't always use it. You can always put a zero here.

Our policy is to get ten hours up front with new clients. If someone is transitioning from break/fix to a signed agreement, and they've been good about paying their bills, we don't require a pre-payment.

As we discussed earlier, many people sign agreements the first time because of a large project. Under these circumstances, they are quite willing to pay a bit up front.

3. Invoices are due in … days. Because we twist arms with our suppliers to give us 30 days, we give our clients 20. Almost everyone pays in the 10-20-day range. A few drag it out longer. Depending on financing, frequency of invoicing, etc. you may be more comfortable with 10 or 30 days.

I certainly wouldn't go beyond 30 unless your company is called The Bank of _____. See the callout on Clients Who Don't Pay.

4. Pay your bill! If you have people who don't pay their invoices, you need to take care of this immediately. The first thing to do is to add those interest charges and do not back down. They get this with their Visa and their American Express, and even with their landlord.

Make this number as high and painful as you can. Make sure your client understands that paying late is not okay. In some states, you can only charge a limited rate unless the client agrees to a higher rate. That's why you need this here. You don't want to find out after it's too late that you can only charge a small sum.

5. What are your hourly minimums? As with everything else, the more consistent you are between clients, the better. For example, you might have a one-hour minimum for onsite and a half hour minimum for remote.

In our case, we cut it to one quarter hour minimum for remote because we have excellent tools for remote support. If you have the right tools, you can do an amazing number of tasks remotely. And if you don't have to pay your staff for travel time to and fro, you can actually make more money with the .25 remote option.

Do your own math. Consider your overall pricing structure and determine what's best here. And make it as consistent as possible!

Call out: Clients Who Don't Pay

I don't know what it is, but there's something about small service companies (like SMB consultants) that leads them to adopt very lax rules around collecting money.

Do not let this slide. Here's what happens when you do: you begin to discount your work. Let's say you bill $1,000 worth of labor. But the client doesn't pay. Eventually, you send it to collections, or negotiate with the client. In either case you've wasted lots more of your time and accepted fifty or seventy-five cents on the dollar.

Stop it! And if you haven't made this mistake, don't.

Somehow, many business professionals feel exempt from the "Net 20 Terms" they agreed to. I've had two examples. Neither one is my client anymore.

Example One: Client simply never paid his bill. We sent a reminder, then an invoice with finance charges. Finally, we sent a note that we don't do work on accounts with past-due balances.

Eventually, of course, he needed further work. We told him that we'd come if he paid us in full plus an additional one hour up front in cash. We showed up, accepted a cash payment, and addressed his immediate problem. Then we sent a nice letter telling him that he needs to find someone else to provide his technical support.

Example Two: Client chronically late. We had another client who always paid her bill, but always paid it late. Many office managers take a class on accounting and learn to pay bills late.

To them, Net 30 meant Due in 30, Late in 60. So, they paid their bill on day 59. And they didn't want any service charges. They wanted us to forgive them because they were "getting back on track." Then they paid one bill on time and went back to 45, 60, 90 days.

Fire these people. Put all that time and energy into something else.

If you simply start out with zero tolerance in this area, you won't have problems. When I talk to consultants who have problems in this area, it's because they transitioned from "no rules" to having rules, or they applied the rules inconsistently.

Concluding Comments

The most important thing about your service agreements is that they exist. In a perfect world, neither you nor your client will ever have to read through them when dealing with an issue that comes up.

You want your business to be friendly and to work as smoothly as possible. But you never know when something can go wrong. Recall Chapters One and Two. There are some very complicated inter-relationships going on here. The longer you stay in business, the bigger and more complicated these relationships become.

Service Agreements literally define your company and your relationships with clients. They define your products, how they're priced, and how they'll be delivered.

But the most important thing that you get from a good service agreement is consistency. You have written "rules" to govern how you do things and what your "policies" are.

If you find that you do something not covered by this sample language, write your own. You're going to have the whole thing reviewed by a lawyer anyway. So throw your change in there and let the attorney tweak it as needed. See Chapter Seven on how to make the most of your attorney.

The most important thing is to have some kind of agreement that governs all of your client relationships. Just do it.

Chapter Appendix:
Software Development Parts

If you write code for clients, you'll need to provide some additional agreements regarding ownership and maintenance of code. These sections are not included in the sample agreements in the next few chapters. They are provided separate from the service agreements on the accompanying downloadable contents

Ownership of Code

There are a few more considerations if you are writing code for clients. These are particularly important if the code is commercially valuable. Whatever you do, make sure you deal with this question up front.

I present three options here. Alternative One is a flat fee for programming and the developer retains rights to the code. The client receives a perpetual, non-exclusive license to use the code.

Alternative Two is basically the same, but adds royalties. Note that Alternative Two is a bit more complicated because you have to deal with royalty payments.

Alternative Three provides the whole code, along with all rights, to the client. You would expect this to be at a higher price.

Ownership of Software:

Alternative One

```
x. Ownership of Software. Developer shall
retain all copyright, patent, trade secret and
other intellectual property rights Developer
may have in anything created or developed
by Developer for Client under this Agreement
("Work Product"). Developer grants Client
a nontransferable license to use the Work
Product. The license is conditioned upon full
payment of the compensation due Developer
under this Agreement.

The license shall be exclusive in the United
States for a period of _____ years
following acceptance by Client of the Software
as set forth in this Agreement. The license
shall automatically revert to a perpetual
nonexclusive license following the period of
exclusivity.

The license shall authorize Client to:
```

(a) install the Software on computer systems owned, leased or otherwise controlled by Client;

(b) utilize the Software for its internal data processing purposes (but not for time-sharing or service bureau purposes); and

(c) copy the Software only as necessary to exercise the rights granted in this Agreement.

Ownership of Software:

Alternative Two

x. Ownership of Software. Developer assigns to Client its entire right, title and interest in anything created or developed by Developer for Client under this Agreement ("Work Product") including all patents, copyrights, trade secrets and other proprietary rights. This assignment is conditioned upon full payment of the compensation due Developer under this Agreement, including royalty fees.

Client agrees to pay to Developer a royalty fee based on the gross revenues generated from the resale of the Work Product. This includes all sales, rental, time-sharing, or service bureau purposes for which Client receives compensation for the use of the Work Product.

For a period of _____ years following acceptance by Client of the Software as set forth in this Agreement, Client shall pay to Developer a royalty fee equivalent to

_____ percent (_____%) of gross revenues generated from the resale of the Work Product. For the next _____ years, Client shall pay to Developer a royalty fee equivalent to ____ percent (_____%) of gross revenues generated from the resale of the Work Product. After the end of the second _____-year period, and continuing in perpetuity, Client shall pay to Developer a royalty fee equivalent to _____ percent (_____%) of gross revenues generated from the resale of the Work Product.

Each calendar quarter, Client shall submit to Developer a Royalty Report specifying the revenues generated from the Work Product through sales, rental, time-sharing, or service bureau purposes. This Royalty Report is due every quarter even if there are no sales. Each Royalty Report will cover one calendar quarter and will be due _____ days after the end of each calendar quarter. Any Royalty fees due to Developer shall be due _____ days after the end of each calendar quarter. The Parties will agree on the form of the Royalty Report.

Developer shall execute and aid in the preparation of any documents necessary to secure any copyright, patent, or other intellectual property rights in the Work Product.

Client grants to Developer a nonexclusive, irrevocable license to use the Work Product. Developer may not sell the Work Product or any portion of the work product.

Ownership of Software:

Alternative Three

x. Ownership of Software. Developer assigns to Client its entire right, title and interest in anything created or developed by Developer for Client under this Agreement ("Work Product") including all patents, copyrights, trade secrets and other proprietary rights. This assignment is conditioned upon full payment of the compensation due Developer under this Agreement.

Developer shall execute and aid in the preparation of any documents necessary to secure any copyright, patent, or other intellectual property rights in the Work Product.

Client grants to Developer a nonexclusive, irrevocable license to use the Work Product. Developer may not sell the Work Product or any portion of the work product.

Maintenance of Code

As a software developer, you will also have to address the issue of code maintenance (fixing bugs). Here's one way to handle that:

x. Maintenance of Software. Beginning on the date of final payment for this Agreement for Software Development, and ending 12 months after that date, Developer shall provide the following error-correction and support services:

(a) telephone hot-line support during Developer's normal days and hours of business operation. Such support shall include consultation on the operation and utilization of the Software. Client shall be responsible for all telephone equipment and communication charges related to such support; and

(b) error correction services, consisting of Developer using all reasonable efforts to design, code and implement programming changes to the Software, and modifications to the documentation, to correct reproducible errors therein so that the Software is brought into substantial conformance with the Specifications.

Payment for Maintenance: At the end of the twelvemonth maintenance period, Client may agree to pay Developer for error-correction and support services the annual sum of _____ dollars ($_____.00), payable on the first day of the first month following expiration of any warranty period. Three years after the date of Client's final acceptance of the Software, Developer shall be entitled to increases in the maintenance fee upon at least _____ days' prior written notice to Client.

If Client chooses not to enter into a Maintenance agreement, or fails to pay for Maintenance, all such Maintenance will be provided at the Developer's current market rate for programming services.

Client's Role in Maintenance: The provision of the error-correction and support services described above shall be expressly contingent upon Client promptly reporting any errors in the Software or related documentation to Developer in writing and not modifying the Software without Developer's written consent.

Maintenance under this Section shall not include any programming that changes the functionality of the Software or adds any features to the Software. Nor shall maintenance include training beyond that provided during the development stages defined in the Development Agreement.

Background Technologies

And, finally, software developers have to make sure that they follow the rules regarding the development tools they use. For example, you can't resell C# development tools. You can embed routines in software, but you don't own the rights to these and you can't sell them.

Sometimes this is a little difficult to explain to a client. They want to "own" the software. You have to explain that this is non-negotiable. You can't sell what you don't own.

x. Ownership of Background Technology. Client acknowledges that Developer owns or holds a license to use and sublicense various preexisting development tools, routines, subroutines and other programs, data and materials that Developer may include in the Software developed under this Agreement. This material shall be referred to as "Background Technology."

Developer retains all right, title and interest, including all copyright, patent rights and trade secret rights in the Background Technology. Subject to full payment of the consulting fees due under this Agreement, Developer grants Client a nonexclusive, perpetual worldwide license to use the Background Technology in the Software developed for and delivered to Client under this Agreement, and all updates and revisions thereto. However, Client shall make no other commercial use of the Background Technology without Developer's written consent.

"Man is an animal that makes bargains; no other animal does this—one dog does not change a bone with another."
— Adam Smith

Chapter Four – Boilerplate: A Sample Credit Agreement for Clients Without a Service Agreement

As I mentioned in Chapter Two, we just plain don't do work without a signature that says we'll get paid. But aside from that, our basic "one-pager" lays out some rules of engagement.

Some things on this list may never be an issue with you. If that's the case, remove them. If you have other concerns or restrictions, add them. The most important thing is to make sure you get paid. The fewer rules it takes to get that done, the better.

At the beginning of Chapter Two I told you to gather up all your policies from invoices, flyers, web pages, etc. This Credit Agreement is the place to put all that stuff.

I do not go through each paragraph and explain it because I think all of these items are truly self-explanatory. This entire agreement is in the downloadable content as Chapter Four Sample Agreement. docx.

The Text:

```
Credit Agreement
Thank you for your business. We look forward
```

to helping you today and for many years to come. As with any relationship, we believe things work more smoothly when we agree on how business will be conducted. Please complete the following agreement.

I/we agree to comply with the following terms:

1. All labor charges are non-refundable.
2. All hardware is sold with a manufacturer's warranty. (Consultant) provides no additional warranty.

Or

If the manufacturer's warranty is less than 1 Year, (Consultant) warranties the hardware from the end of the manufacturer's warranty to the end of 1 Year from the date of purchase.

3. All merchandise may be returned within 30 days. There is a 10% restocking fee on all items except special order items. There is a 20% restocking fee on special order items.

4. Licensed software is not refundable.

5. There is a $25 charge on returned checks.

6. Invoices are due and payable to (Consultant) upon completion of the work.

7. All unpaid sums that are not in dispute shall bear interest at the rate of 1.5% per month.

8. Cost of collection, including reasonable

attorney's fees, shall be borne by the client.

9. All merchandise remains the property of (Consultant) until paid in full.

10. (Consultant) shall not be bound by any terms or conditions printed on a purchase order, check, or correspondence from client without prior written acceptance of such terms.

11. Quotations and responses to requests for quotations do not include the price of sales tax or shipping unless these items are explicitly stated. Client is responsible for the cost of sales tax and shipping of all merchandise.

12. (Consultant) will not perform for clients with past due balances. This includes but is not limited to emergency services.

13. From time to time (Consultant) may offer for sale items that are demonstration units, refurbished, or used. All such items are sold AS-IS and are not returnable.

14. (Consultant) does not guarantee the price and / or the availability of product and / or services quoted.

15. All defective merchandise must be returned to (Consultant), prepaid.

LIMITATION OF LIABILITY

16. Consultant warrants to Client that the

material, analysis, data, programs, and services to be delivered or rendered under this Agreement will be of the kind and quality designated and will be performed by qualified personnel.

Consultant offers no guarantees or warranties, express or implied, as to system availability and functionality during any phase of its support services and makes no guarantees or warranties, expressed or implied, regarding the ability to resolve computer-related problems, to recover data, or to avoid losing data.

Consultant makes no other warranties, whether written, oral or implied, including without limitation warranty of fitness for purpose of merchantability. In no event shall Consultant be liable for special or consequential damages, either in contract or tort, whether or not the possibility of such damages has been disclosed to Consultant in advance or could have been reasonably foreseen by Consultant, and in the event this limitation of damages is held unenforceable then the parties agree that by reason of the difficulty in foreseeing possible damages all liability to Client shall be limited to _____ ($____.00) as liquidated damages and not as a penalty.

LEGAL ACTION AND JURISDICTION

17. Jurisdiction. The enforcement of this agreement shall be governed by the laws of the State of _____.

I/We agree to these terms and accept responsibility for payment of our account.

Client Signature _____

Date _____

Print Name _____

Consultant Signature _____

Date _____

Print Name _____

Discussion:

Depending on the font size used, this agreement will either be one page, one side or one page, two sides. As a rule, don't make the font too small or people will think you're trying to get away with something.

Our procedure is to print these forms back-to-back and make sure each technician has a few with him at all times. We have every new client sign this form and then we leave a copy with the client.

We also print this up as a .pdf document and put it on our SharePoint intranet so that our technicians have access to it at all times.

Most people don't read this agreement. But they all sign it! We don't think there's anything here that's awkward, cumbersome, or difficult. It's just business.

"I think we may safely trust a good deal more than we do."
— Henry David Thoreau

Chapter Five – A Sample Service Agreement for Hourly Services

Introductory Comments

Having discussed all the "parts" in Chapter Three, this chapter simply presents a complete sample agreement without the commentary.

We put this into a Word file with our headers and footers, logo, etc. We print up two copies and sign both of them. Then the client signs both and we each keep one copy.

If you choose to have agreements that expire, be sure to enter the beginning and ending dates into your professional services automation (PSA) tool, customer relationship management (CRM) system, spreadsheet, or whatever you use.

This entire agreement is in the downloadable content as Chapter Five Sample Agreement.docx.

The Text:

```
This  Agreement  is  made  as  of  _____,
_____  between  _____  ("Consultant")  and
_____  ("Client").

1.  Term  of  Agreement.  This  Agreement  for
```

Consulting Services is for the Period _____ to _____. This agreement may be extended by execution of an extension by both parties. Either party may cancel this agreement upon thirty (30) days written notice.

Or:

1. Term of Agreement. This agreement shall commence on the date set forth above and will continue until either party hereto provides the other with written notice upon thirty calendar days' notice of termination.

2. Early Termination. Client recognizes that the preferred rate defined in this Service Agreement is contingent upon the purchase of a specified number of labor hours. In the event that Client terminates this Agreement, for any reason other than failure to perform the services outlined herein, before the Client has purchased the number of hours agreed upon, Client agrees to pay an early termination fee of $_____.

3. Complete Agreement. This Agreement constitutes the entire Agreement between Consultant and Client. No other representations, whether written or oral, shall bind the parties with regard to the matters agreed to in this Agreement. Consultant and Client acknowledge that they are executing this Agreement based solely on the written text of this Agreement.

4. In the event of a conflict between the

text of this Agreement and the text of any Attachments to this Agreement, the provisions of the attachments will prevail.

5. Jurisdiction. The enforcement of this agreement shall be governed by the laws of the State of _____.

6. Enforcement. If any piece of this agreement is held to be unenforceable or not fully enforceable, then the parties agree that the remainder of this Agreement will remain enforceable and unaffected. It is the intent of the parties that this Agreement, and each of its parts, will remain in effect to the extent that they not affected by legal or judicial findings.

7. Dispute Resolution. Should dispute arise with regard to the enforcement or application of any provisions of this Agreement, the parties agree to binding arbitration by a mediator associated with the American Arbitration Association.

The mediator's arbitration decision will be enforceable according to the laws of the State of _____.

8. Notices.

(i) Notices to Consultant should be sent to:

_____

_____

_____

_____

(ii) Notices to Client should be sent to:

_____

_____

_____

_____

9. Services Provided. Consultant will provide the following Services under this agreement:

Item a. Consultant agrees to provide consulting services related to computer hardware, software, network configuration, operating systems and networks, database development, programming, and other similar work. Such work will be done on behalf of Client and will be performed at a location or locations to be determined by Consultant. For example, services might include installing and configuring Windows operating systems, Internet servers, and associated software.

Item b. Regular Consulting Hours.

Regular Consulting Hours shall consist of any time Consultant works for Client during regular business hours. Regular business hours are defined as 8:00 AM to 5:00 PM Monday through Friday excluding national holidays.

Item c. Extended Consulting Hours.

Extended Consulting Hours shall consist of any hours worked by Consultant outside of the period defined as "Regular" Consulting Hours. This includes weekends, the period 5:00 PM to 8:00 AM during weekdays, and all holidays.

Item d. Programming Hours.

Programming Hours shall consist of any work by Consultant using a programming language such as (but not limited to) PHP, HTML, Visual Basic, VBScript, C/C++, or Java.

10. Additional Work. Upon request from Client to perform services, whether specifically listed in this agreement or not, Consultant is authorized to perform such services and Client agrees to pay Consultant for such services at the rates specified in this Agreement.

11. Cost of Services.

All General Consulting Hours, as defined in this Agreement will be provided at a rate of _____ dollars ($_____) per hour.

All Extended Consulting Hours, as defined in this Agreement will be provided at a rate of _____ dollars ($_____) per hour.

All Programming Hours, as defined in this Agreement will be provided at a rate of _____ dollars ($_____) per hour.

These rates are subject to change, subject to a thirty (30) day written notice.

12. Confidential Information. Each Party hereto ("Such Party") shall hold in trust for the other Party ("Such Other Party"), and shall not disclose to any non-party to the Agreement, any confidential information of Such Other Party. Confidential information

is information which relates to Such Other Party's research, development, trade secrets, or business affairs, but does not include information which is generally known or easily ascertainable by nonparties of ordinary skill in computer design, programming, networking, information technology, or the specific business interests of either Party.

13. Staff. Consultant is an independent Contractor and Consultant is not employed by Client. Consultant agrees to provide services as a contractor under this Agreement. Consultant will determine which services are to be performed, and the order and method of providing those services in response to requests from Client.

Consultant provides services to a variety of clients and is not required to provide full time availability for services to client. Consultant agrees to pay all taxes related to the work performed under this Agreement. Client is not providing any insurance coverage on the Consultant or Consultant's staff.

14. Non-Solicitation of Employees. During the term of this agreement and for a period of twelve (12) months after the termination of this agreement for any reason, Client agrees not to recruit or employ any employee of Consultant without the prior written consent of the President or Chief Executive Officer of Consultant. Consultant hereby agrees that it will not recruit or employ any of Client's employees without written consent from Client.

15. Drug Free Workplace Certification. Consultant agrees to comply with the provisions of (State Law or regulatory section) regarding maintenance of a Drug-Free Workplace. Consultant agrees to notify its employees that they are prohibited from engaging in the unlawful manufacture, distribution, dispensation, possession or use of controlled substances.

LIMITED WARRANTY

16. Liability. Consultant warrants to Client that the services and labor performed as part of this Agreement will be of manner and quality associated with a professional technical consultant.

Consultant offers no guarantees or warranties as to system availability and functionality during any phase of its support services and makes no guarantees or warranties regarding the ability to resolve computer-related problems, to recover data, to avoid losing data, or to prevent loss of income.

Consultant makes no other warranties that products delivered under this Agreement are fit for merchantability. In no event shall Consultant be liable to the client for any loss of profit, loss of business, consequential, or any other damages as a result of work performed under this Agreement. In the event that this limitation of damages is held unenforceable then the parties agree that all liability to Client shall be limited to _____ dollars ($_____.00).

Either:

all liability to Client shall be limited to
_____ dollars ($_____ .00).

Or:

. . . all liability to Client shall be limited
to the total dollar amount of services paid
by Client to Consultant in the previous three
(3) full months.

17. Terms.

a) Client agrees to purchase a minimum of
_____ hours of Services during the Term
of this Agreement.

b) Client agrees to pre-pay _____ hours of
labor upon execution of this agreement.

c) All invoices to Client shall be due within
_____ days.

d) Any unpaid sums over _____ days old
that are not in dispute shall bear interest
at the rate of _____ percent per month.
Costs of collection including reasonable
attorney's fees shall be borne by the Client.

e) There is a _____ minimum charge for
all onsite visits for services not covered by
a separate managed services agreement. There
is a _____ minimum charge for remote support
services not covered by a separate managed
services agreement.

IN WITNESS WHEREOF, the Parties hereto have signed this Agreement and agree that it shall be binding upon the parties and their respective heirs, successors, and assigns.

Consultant _____

Date_____

(You, your title)

(Your company)

Client_____

Date_____

(Client name, title)

(Client company)

"We are confronted with insurmountable opportunities."
— Pogo

Chapter Six – Flat Fee or Managed Services Parts and Sample Agreement

Introductory Comments

We execute the agreement for managed services as an attachment or amendment to our normal service agreement. Notice that this actually represents a statement of our operating procedure. We only offer managed services to clients with a service agreement.

Recall from Chapter Two that we use a multi-tiered approach. We put the regular service agreement in place to cover all hourly labor. Then we add the managed service agreement onto that. The result is that any labor not covered by the flat-fee agreement will be charged according the underlying service agreement.

To be honest, this is mostly a fluke of circumstances. We've signed service agreements and maintenance agreements for ten years. So when we added flat-rate pricing, it made sense to do so as an attachment or amendment.

Having said that, I would do it the same way if I were starting over today. Since we sign more basic service agreements than we sign managed service attachments, I don't want to create one big contract with a whole section that gets crossed out.

At the same time, I don't want to create a "combined" contract and have to worry about keeping changes synchronized between the service agreement and the service agreement portion of the managed services agreement. I'm confusing myself just trying to write it out!

If you flip back to Chapter Two, there's a lengthy discussion of how we came up with our managed services. This service agreement represents the plan described there. If you have a "tiered" service with Gold, Silver, and Bronze levels, you'll need to add a paragraph describing these and then have a checkbox to indicate what the client is buying.

Again, we put this in a nice Word file with our header, logo, etc. Fill out the totals, sign two copies, and give one to the client. That's all there is to it!

The Managed Services "Parts"

The Text

```
Agreement for Managed Services
(Attachment to Agreement for Consulting
Services)

This Agreement for Managed Services (this
"Attachment") is made as of _____ by and
between _____ ("Consultant") and _____
("Client"). This Attachment amends and
modifies, and constitutes an "Attachment" to,
that certain Agreement for Consulting Services
previously executed by Consultant and Client
(the "Agreement"). All of the terms of the
Agreement are expressly incorporated herein.
All capitalized terms herein shall have the
meaning ascribed to them in the Agreement,
unless expressly defined otherwise herein.
```

You should have a lawyer review every agreement or contract you sign.
www.smallbizthoughts.com

Discussion

This is basic housekeeping. We're drafting an attachment to the Service Agreement.

The Text

x. Services; Payment Terms. Consultant agrees to perform for Client the following services (the "Managed Services") for the following monthly fixed fees:

(a) Network Equipment Support
times __1__ at $ __250.00__ = $ ____

(b) Server Support
times _____ at $ __400.00__ = $ ____

(c) Servers with SQL, Exchange, etc.
times _____ at $ __400.00__ = $ _____

(d) Desktop or Laptop Computers Support
times _____ at $ __65.00__ = $ _____

(e) Desktop or Laptop Computers Remote Backup (25GB limit)
times _____ at $ __25.00__ = $ _____

    Monthly Total:$ _____

Discussion

Obviously, this is just a sample. You'll fill in your pricing and your options from your three-tiered price list. Here's a sample of what it might look like:

(a) Network Equipment Support
     times 1 at $ 250.00 = $ 250

(b) Server Support
     times 1 at $ 400.00 = $ 400

(c) Servers with SQL, Exchange, etc.
     times NA at $ 400.00 = $

(d) Desktop or Laptop Computers Support
     times 27 at $ 65.00 = $ 1,755

(e) Desktop or Laptop Computers Remote Backup (25GB limit)
     times NA at $ 25.00 = $
     Monthly Total: **$ 2,405.00**

The Text

In addition to the monthly fees set forth above, Client agrees to pay an initial setup of remote monitoring services ("Setup Services") fee in the amount of $ _____ .

Client shall pay the Setup Services fee (if any) upon execution of this Attachment, and the Managed Services fees on the first day of each calendar month. Note that there are no partial months or prorated fees.

Discussion

We charge a setup fee of 100% of the first month's charges. We sometimes discount this based on whether it's early or late in the month. For the most part, the only cost to us for the first few weeks is the labor to start installing our Remote Monitoring software, which is a few minutes per machine. So, if it helps the client to sign because she thinks she's saving money, we throw in a discount.

I've seen consultants who charge a 25%, 50%, or flat-rate setup fee. In general, I think you should charge something for your effort. I've never had a client question this.

The Text

x. Services Included in Managed Services.

(a) "Network Equipment Support" shall consist of all labor related to maintaining configuration, logging (if possible and appropriate), and monitoring of network equipment, including routers, firewalls, switches, spam filters, and other equipment used to move, monitor, or intentionally affect Ethernet traffic on Client's local area network.

Network Equipment Support shall also consist of working with Client's Internet Service Provider to maintain proper configuration of Internet equipment at Client's office, whether owned by Client or Client's ISP. Consultant will provide all service related to these products.

Discussion

Our support of "the network" is a little unusual. Many consultants don't offer this at all, or they put lots of limitations on it. We charge a fee that seems reasonable (it's in line with the server maintenance), but is actually rather high. If you consider that $250 is about two hours' labor per month, this comes out to 24 hours a year.

Our network maintenance is different from the server and desktop maintenance because it includes labor related to hardware maintenance. In fact, we cover all labor related to moving from one ISP to another. This is potentially huge. After all, almost every client has a horror story of a transition that took ten or more hours of labor.

We cover labor related to an ISP switch because we're playing an insurance game here. After all, how often will a client change ISPs? In our experience, it is less than once every three years. And most of those are not disasters. But we can tell the client "If it takes ten hours, you're totally covered."

I must say, for the record, that we are extremely good with network equipment. We estimate 30-60 minutes to configure virtually any firewall, router, or spam filter because of strict equipment choices and well-documented procedures. If we found ourselves spending hours and hours with every install, we'd have to rethink this offering.

Most networks, and most network equipment, are very stable. On average we spend zero hours per year troubleshooting a given network.

The Text

(b) "Server Support" shall consist of all labor related to maintaining Client's server operating system, any programs included in the operating system, backup software, virus scanning software, hard disk defragmentation software, and the following programs installed after the operating system:

_____

_____

_____

Client agrees that Client will maintain

```
separate service agreements with these software
vendors. Consultant will coordinate or provide
all service related to these products.
```

Discussion
We don't let the list of "additional" programs get too long. We've been bit by that once. But a couple of line of business (LOB) applications are no problem. But we also do not replace the LOB support system.

LOB Policies

LOB or Line of Business applications can be very easy to deal with, or they can be a real pain in the neck. There doesn't seem to be much in the middle.

There are two keys to success with LOBs. First, you — the managed service provider — must be in charge. Second, the client must maintain support contracts for these products.

When you combine these, it means that you keep a list of SQL passwords, support phone numbers, etc. All patches go through you. All updates go through you. That doesn't mean you provide the labor, but it does mean that these providers can't just remote into a system and apply changes without keeping you in the loop.

This process may be a bit labor-intensive at first, but you'll come out ahead in the long run.

And if the client does not pay for maintenance from an LOB provider? Then all of your labor related to that product is billable. You win either way.

Aside from LOB applications, our goal is to provide all the support needed to keep the server in good (great!) working order. The mantra here is "Maximize uptime. Minimize downtime. Ohm, ohm, ohm."

The Text

(c) "Servers with SQL, Exchange, etc." shall consist of all labor related to maintaining the following specific software packages on any of the servers listed in Section x (b) above:

_____

_____

Consultant will provide all services related to these products.

Discussion

Here, list whether the machine is SQL, Exchange, etc. You do not need to list the LOB applications as they were listed above. Remember, this is the $100/month add-on for having a machine with a major package that requires a little extra labor. All I would expect here is "Exchange" or "SQL" or whatever the main product it. Remember every server is listed in item (b) above. As a result, if a SQL server has an LOB, the LOB would be listed in item (b) and the fact that it's SQL would be listed in item (c).

The Text

(d) "Desktop or Laptop Computer Support" shall consist of all labor related to maintaining the computer operating system, any programs included in the operating system, Microsoft Office products, virus scanning software, and the following programs installed after the operating system:

_____

_____

Client agrees that Client will maintain
separate service agreements with these
software vendors. Consultant will provide all
service related to these products.

Discussion

Unless there's some really difficult LOB application, it is our goal
to provide 100% of the support on the desktops. This keeps us in
charge of updates, patches, fixes, etc.
We rely a great deal on remote support whenever possible. Because
we can script updates and get reports about any updates that were
unsuccessful, our labor to maintain systems is minimal.

We do include the first three hours of labor related to setting up a
new system. If you excluded that, you would be able to offer a lower
monthly price.

The Text

(e) "Desktop or Laptop Remote Backup" shall
consist of all labor related to creating a
backup "image" of the client machine on the
_____Your Company____ remote backup systems,
or on a server of the Client's choice. This
service does not include the labor needed to
restore files or systems in case such services
are necessary.

Client agrees to provide and maintain both
a storage device capable of handling this
backup and an Internet connection sufficient to
copy the backup files offsite in a reasonable
amount of time.

Discussion

We don't generally offer backups for desktops. But there are many options today. Some clients still have very small Internet connections, which can make this a challenge. But if you are running something like Server 2016 Essentials, you can easily backup desktop machines to the server and then send those backups to cloud storage.

The Text

```
x. Additional Machines. Client may add or
remove services for additional servers,
desktop PCs, or laptop PCs by opening a service
ticket with Consultant or sending an official
request by email to Consultant. Consultant
agrees to keep an accurate list of machines
covered under this Attachment and to provide
this list to Client upon request. Note that
there is a four-month minimum for all machines
added to Managed Services.
```

Discussion

This section simply means that the client can add or remove machines from the managed service plan as needed. Obviously, it is in your best interest to keep accurate track of this and bill accordingly.

We make it very simple to change the plan. We don't want a hassle when a client adds or removes a machine. And since we generally set up the machines and hook them to the network, we know when one goes in.

I'm quite surprised when I hear consultants say that clients just drop a new machine on their network. I honestly don't understand how that can happen. If you are managing your client's network then you're the one who sets up new machines.

The Text

x. Software Updates. Maintaining the systems described above shall include applying all appropriate software and operating system updates in a reasonable amount of time. Consultant shall determine when software updates are appropriate and what constitutes a reasonable amount of time.

Client acknowledges that if Client requests updates that Consultant considers inappropriate, or wishes to have updates applied before Consultant deems them safe, Consultant is not responsible for the consequences of such actions and Client may be charged a Regular Consulting Hours or Extended Consulting Hours charge, as the case may be, for all labor related to the consequences of such actions.

Furthermore, if Client performs or allows anyone other than Consultant to perform any maintenance on any of these machines, Consultant is not responsible for the consequences of such actions and Client may be charged a Regular Consulting Hours or Extended Consulting Hours charge, as the case may be, for all labor related to the consequences of such actions.

Discussion

This section simply states that we are in charge of the updates. We decide what goes in and when, and we do the work. While this might appear as a territorial move, it's really just common sense. We're going to be responsible for fixing anything that goes wrong. Therefore, we want to be the ones to manage any events that might go wrong.

This is also the meat of what you're providing. Up above we said we'd maintain the machines. Here we state what maintenance means.

Note, also, that this is where we make clear that the client is not to perform any maintenance or other work. If the client messes up something on the server, or a workstation, then we charge by the hour to fix it.

The Text

```
x. Monitoring Software. In order to provide
the services specified in this Attachment,
Consultant must install remote monitoring
and management software on Client's servers,
desktop computers, laptops, and possibly
other equipment at Client's office. Client
grants permission to Consultant to install
remote monitoring and management software from
Manufacturer , or any other remote monitoring
and managing software deemed necessary by
Consultant.
```

Discussion

We started out by using a home-grown collection of free tools and the features built into Windows Server and Small Business Server. Eventually, we bought into a software program that allows us to maintain and manage a large number of machines in a smaller period of time.

As with most "managed services" offerings, there is an agent that has to be installed on the relevant machines. We make sure the client knows this and agrees to it. Note, below, that we also make clear that we'll remove this if they no longer want the service.

The Text

```
x. Term of Attachment; Termination.

(a) This Attachment shall commence on the
date set forth above and shall continue until
the earlier of: (i) thirty (30) calendar days
after either party hereto provides the other
with written notice of termination; or (ii)
the termination of the Agreement.
```

Discussion

We want the managed services agreement to go on forever, but that may not always be the case. So, here we state that the client can back out of the MSA with thirty days' notice. As I mentioned earlier, we make it very easy for the client to get out. This is a selling point.

Recall that we execute the MSA as an attachment to the standard agreement. So, if they cancel the Agreement for any reason, the MSA would also be cancelled.

The Text

```
(b) Upon termination of this Attachment,
Consultant shall uninstall all remote
monitoring and management software from
all Client equipment. Client acknowledges
that this may leave its computers and other
equipment without adequate systems for
updates to operating systems, software, and
```

```
virus scanning programs. Consultant shall
not be held responsible for any damages or
consequences resulting from the removal of
remote monitoring and management software.
```

Discussion

If client chooses to cancel an agreement, we need to recoup our agent license. Obviously, we must therefore remove the agent. The result may be that a machine is left with Windows Updates disabled, as well as not monitoring of virus scanning, file defragmentation, success of backups, etc.

We don't intend this paragraph to be a penalty of any kind. It simply represents the truth of the situation. After all, when a client moves from a "managed" environment to an un-managed environment, one would expect all the things that had been managed to now be done manually — or not at all.

The Text

```
x. Nature of This Attachment. This Attachment
is intended to cover the maintenance of
computer operating systems and software only.
It is not intended to cover any hardware,
materials, equipment, consumables, hardware
failures, troubleshooting, replacements, or
any labor related to projects other than the
proper maintenance of operating systems and
software. Consultant offers other services,
including hardware-related labor. Any labor
provided outside the scope of this Attachment
will be at the rates stated in the Agreement.
```

Discussion

This paragraph is thrown in to make clear the intention of the MSA. We want to make sure the client understands that he's not getting a blanket agreement for all labor related to anything he can think of.

This is a Maintenance Agreement. So, we cover everything related to general maintenance and charge extra for everything else.

Note: This may be the most important paragraph in the service agreement. At a minimum, it's the paragraph that will keep you profitable. There's no such thing as "All you can eat" and the managed service agreement excludes Adds, Moves, and Changes.

The Text

```
IN WITNESS WHEREOF, the parties hereto have
executed this Agreement as of the date first
above written.
Consultant: _____
Name:_____
Title:_____

Client:_____
Name:_____
Title:_____
```

Discussion

Blah blah blah. Execute the Attachment. Same as with the underlying Agreement, both parties sign two copies. Each party keeps one.

Sample Managed Services Agreement

Those were the Parts. The following is a copy of the entire MSA Attachment:

Agreement for Managed Services

(Attachment to Agreement for Consulting Services)

This Agreement for Managed Services (this "Attachment") is made as of _____ by and between _____ ("Consultant") and _____ ("Client"). This Attachment amends and modifies, and constitutes an "Attachment" to, that certain Agreement for Consulting Services previously executed by Consultant and Client (the "Agreement"). All of the terms of the Agreement are expressly incorporated herein. All capitalized terms herein shall have the meaning ascribed to them in the Agreement, unless expressly defined otherwise herein.

1. Services; Payment Terms. Consultant agrees to perform for Client the following services (the "Managed Services") for the following monthly fixed fees:

(a) Network Equipment Support
times __1__ at $ 250.00 = $ _____

(b) Server Support
times _____ at $ 400.00 = $ _____

(c) Servers with SQL, Exchange, etc.
times _____ at $ 400.00 = $ _____

(d) Desktop or Laptop Computers Support
times _____ at $ 65.00 = $ _____

(e) Desktop or Laptop Computers Remote Backup

(25GB limit)
times _____ at $ 25.00 = $ _____
   Monthly Total:$_____

In addition to the monthly fees set forth above, Client agrees to pay an initial setup of remote monitoring services ("Setup Services") fee in the amount of $ _____ .

Client shall pay the Setup Services fee (if any) upon execution of this Attachment, and the Managed Services fees on the first day of each calendar month. Note that there are no partial months or prorated fees.

2. Services Included in Managed Services.

(a) "Network Equipment Support" shall consist of all labor related to maintaining configuration, logging (if possible and appropriate), and monitoring of network equipment, including routers, firewalls, switches, spam filters, and other equipment used to move, monitor, or intentionally affect Ethernet traffic on Client's local area network.

Network Equipment Support shall also consist of working with Client's Internet service provider to maintain proper configuration of Internet equipment at Client's office, whether owned by Client or Client's ISP. Consultant will provide all service related to these products.

(b) "Server Support" shall consist of all labor related to maintaining Client's server operating system, any programs included in the operating system, backup software, virus scanning software, hard disk defragmentation software, and the following programs installed after the operating system:

_____

_____

Client agrees that Client will maintain separate service agreements with these software vendors. Consultant will coordinate or provide all service related to these products.

(c) "Servers with SQL, Exchange, or SBS Support" shall consist of all labor related to maintaining the following specific software packages on any of the servers listed in Section x (b) above:

_____

_____

Consultant will provide all service related to these products.

(d) "Desktop or Laptop Computer Support" shall consist of all labor related to maintaining the computer operating system, any programs included in the operating system, Microsoft Office products, virus scanning software, and the following programs installed after the operating system:

_____

_____

Client agrees that Client will maintain separate service agreements with these software vendors. Consultant will provide all service related to these products.

(e) "Desktop or Laptop Remote Backup" shall consist of all labor related to creating a backup "image" of the client machine on the Your Company remote backup systems, or on a server of the Client's choice. This service does not include the labor needed to restore files or systems in case such services are necessary.

Client agrees to provide and maintain both a storage device capable of handling this backup and an Internet connection sufficient to copy the backup files offsite in a reasonable amount of time.

3. Additional Machines. Client may add or remove services for additional servers, desktop PCs, or laptop PCs by opening a service ticket with Consultant or sending an official request by email to Consultant. Consultant agrees to keep an accurate list of machines covered under this Attachment and to provide this list to Client upon request. Note that there is a four-month minimum for all machines added to Managed Services.

4. Software Updates. Maintaining the systems described above shall include applying all appropriate software and operating system updates in a reasonable amount of time. Consultant shall determine when software updates are appropriate and what constitutes a reasonable amount of time.

Client acknowledges that if Client requests updates that Consultant considers inappropriate, or wishes to have updates applied before Consultant deems them safe, Consultant is not responsible for the consequences of such actions and Client may be charged a Regular Consulting

Hours or Extended Consulting Hours charge, as the case may be, for all labor related to the consequences of such actions.

Furthermore, if Client performs or allows anyone other than Consultant to perform any maintenance on any of these machines, Consultant is not responsible for the consequences of such actions and Client may be charged a Regular Consulting Hours or Extended Consulting Hours charge, as the case may be, for all labor related to the consequences of such actions.

5. Monitoring Software. In order to provide the services specified in this Attachment, Consultant must install remote monitoring and management software on Client's servers, desktop computers, laptops, and possibly other equipment at Client's office. Client grants permission to Consultant to install remote monitoring and management software from Manufacturer, or any other remote monitoring and managing software deemed necessary by Consultant.

6. Term of Attachment; Termination.

(a) This Attachment shall commence on the date set forth above and shall continue until the earlier of: (i) thirty (30) calendar days after either party hereto provides the other with written notice of termination; or (ii) the termination of the Agreement.

(b) Upon termination of this Attachment, Consultant shall uninstall all remote monitoring and management software from all Client equipment. Client acknowledges that this may leave its computers and other equipment without adequate systems for updates to operating systems, software, and virus scanning programs. Consultant shall not be held responsible for any damages or consequences resulting from the removal of remote monitoring and management software.

7. Nature of This Attachment. This Attachment is intended to cover the maintenance of computer operating systems and software only. It is not intended to cover any hardware, materials, equipment, consumables, hardware failures, troubleshooting or replacements, or any labor related to projects other than the proper maintenance of operating systems and software. Consultant offers other services, including hardware-related labor. Any labor provided outside the scope of this Attachment will be at the rates stated in the Agreement.

IN WITNESS WHEREOF, the parties hereto have executed this Agreement as of the date first above written.

Consultant: _____
Name:_____
Title:_____

```
Client:_____
Name:_____
Title:_____
```

Memo on Tiered Offerings

Readers of the first edition of this book will know that I've changed my mind on tiered offerings. This comes from two important perspectives. First, my personal experience. And, second, massive amounts of research into the psychological effects of tiered pricing.

I had great luck with a one-size-fits-all offering. I only the sold the best, so take it or leave it. But I also got some pushback from clients who are super happy with the way things were (break/fix). In my book *Managed Services in a Month*, I tell the story of developing my three-tiered offering, so I won't repeat it here. Suffice it to say, when I made the switch to offering tiers, it was much easier to sell the highest level of managed services.

As I say in Managed Services in a Month:

> There's something magical and simple about three options. Some people want the "best," whatever it is. Some want the cheapest. People don't like to be sold, but they love to shop. With three tiers, they can pick the one they want. Depending on how you structure it, the option they pick will reflect where they put the focus on technical support.

The exercise of developing a 3-tiered offering is very easy. First, you create a table in Word. Make columns for your offerings and one to describe the services included.

The default offerings seem to be Silver, Gold, and Platinum. Some consultants use additional tiers with Diamond, Platinum, and other stuff. You can also use names such as Basic, Monitoring, or Complete Care.

The table on the next page is available in the downloadable content that accompanies the book:

Sample – Server Management – Tiered Options

Services Included	Bronze	Silver	Gold
Monitoring	X	X	X
Reporting	X	X	X
Monthly Maintenance	X	X	X
Anti-Virus Monitoring and Maintenance	X	X	X
Patching		X	X
Backup Monitoring and Maintenance		X	X
Support Response Time	6 hr	4 hr	2 hr
Server Operating System Updates			X
Server Software Updates			X
Price per Server per Month	$150	$250	$350

This sample is based on a "flat fee" model for maintenance. Note: NOT All You Can Eat. If you go with a block-of-time model, you would list the number of hours included. The problem with the block-of-time model is that clients tend to keep track of the "unused" hours.

This is also a very short list of services included. You can make this list as long and complicated as you wish. Other items to consider are:

- Quarterly Check-up and Strategy Meeting
- Spam filtering
- Assigned support technician or account manager
- X hours of remote support

You should have a lawyer review every agreement or contract you sign.
www.smallbizthoughts.com

- Y hours of onsite support
- Coordinate service with Line-of-Business software vendors
- Monitor network equipment
- Log analysis for firewalls and switches
- IntraNet maintenance
- Updates to primary software packages (SQL, Exchange, etc.)
- Monthly security reports
- Monthly monitoring reports
- Client access to all service call history
- Easy creation of service calls via web interface
- Coordinate all interaction with hardware vendors for warranty service
- Monitor hard drive fragmentation and general storage system health
- Monthly reports on Internet bandwidth availability

Don't get carried away and give away the store. Having said that, if you have a system in place that allows you to do monitoring and generate many of these reports automatically, then throw it in! If you put together a tiered offering, make sure you do all the math calculations similar to the exercises in Chapter Two. Again, you don't want to create a system that loses money.

Perhaps the best way to approach pricing is to look at the list of services and estimate the average cost to provide these services per month. Some months will take more labor and some less. But in the long run, with lots of clients and lots of servers, the plan needs to make money—enough money to make it worth your while.

As with anything else in your business, you need to create a system that works for you and seems logical and comfortable for you. Over the last fifteen years there has been an explosion of interest in managed services. While many SMB consultants were quick to jump on this train, many others are still not convinced it's right for them or their clients.

Some consultants have no managed service offering and don't

intend to create one. They have a business model that works based on hourly charges. They work with a clientele that understands and is comfortable with hourly charges. As a result, they see no reason to change.

While I think everyone should perform work with some kind of signed agreement in place, that doesn't mean you need to have a "managed services" or flat-rate price structure. Play with the numbers. Consider your options. And . . .

Seek the help of a financial professional. For more tips on that, turn to the final section of this book.

"The main thing is to keep the main thing the main thing."
— Stephen Covey

Chapter Seven – Sample Managed Services Agreement
(Not executed as an amendment)

I really like the idea of signing a basic agreement for consulting and executing the "managed" piece as an add-on. But in the last managed service business, I was asked to combine these into a single contract. So that's what I'm posting here. You've seen all the "parts" before so I won't go over that again.

I've underlined but no anonymized the parts unique to my company and this client, just so you can see how it looks in real life. Sometimes all the underlined blank spaces don't make sense.

This text is in your downloadable content as Chapter Seven Sample Managed Services Agreement.docx.

I think this example is good for illustrating some of the key elements that keep you profitable:

Section One: Defined managed services and spells out that adds-moves-changes are not included.

Section Two: There are no partial months and all machines on managed service must be covered for a minimum of four months. Also, if you terminate services, your machines may be unsafe and not virus scanned.

Section Thirteen: Some good details about what's covered and how service tickets are worked.

Section Fifteen: Getting paid in advance. Other good finance policies.

Section Seventeen: Sample of per-user pricing.

Agreement for Consulting Services

This Agreement is made as of _____ between <u>Great Little Book Publishing Co., Inc., a California Corporation doing business as Small Biz Thoughts</u> ("Consultant") and <u>_____ Client _____</u> ("Client").

In the event of a conflict in the provisions of any attachments hereto and the provisions set forth in this Agreement, the provisions of such attachments shall govern.

1. **Services.** Consultant agrees to perform for Client the services listed in the Section 12 "Scope of Services," below. Such services are hereinafter referred to as "Services."

The service known as "Managed Services" is intended to cover the maintenance of computer operating systems and software only. It is not intended to cover any materials, equipment, consumables, hardware failures, troubleshooting or replacements, or any labor related to projects other than the proper maintenance of operating systems and software. Consultant offers other services, including hardware-related labor. Any labor provided outside the scope of this Agreement will be at the rates stated in the Agreement.

This Agreement does not cover "Adds, Moves, or Changes" to operating systems, software, hardware, or user configurations. This is a maintenance agreement.

2. **Term of Agreement.** This agreement shall commence on the date set forth above and shall continue until thirty (30) calendar days after either party hereto provides the other with written notice of termination. All managed services payments are due on the first of the month and are for the whole month. There are no partial or prorated months. So, for example, if notice of termination is given on the 15th of the month, the agreement will be in effect until the end of the next month.

Client recognizes that the preferred rate defined in this Service Agreement is contingent upon the purchase of a specified number of labor hours. In the event that Client terminates this Agreement, for any reason other than failure to perform the services outlined herein, before the end of the fourth month, client agrees to pay the amount that would be due for the first four months of service.

For any machine added to this Managed Service Agreement there is a four-month minimum for service.

Upon termination of this Agreement, Consultant shall uninstall all remote monitoring and management software from all Client equipment. Client acknowledges that this may leave its computers and other equipment without adequate systems for updates to operating systems, software, and virus scanning programs. Consultant shall not be held responsible for any damages or consequences resulting from the removal of remote monitoring and management software.

3. **Confidential Information.** Each Party hereto ("Such Party") shall hold in trust for the other Party ("Such Other Party"), and shall not disclose to any non-party to the Agreement, any confidential information of Such Other Party. Confidential information is information which relates to Such Other Party's research, development, trade secrets, or business affairs, but does not include information which is generally known or easily ascertainable by nonparties of ordinary skill in computer design, programming, networking, information technology, or the specific business interests of either Party.

Consultant hereby acknowledges that during the performance of this contract, the Consultant may learn or receive confidential Client information and therefore Consultant hereby confirms that all such information relating to the Client's business will be kept confidential by the Consultant, except to the extent that such information is required to be divulged to the Consultant's clerical or support staff or associates in order to enable Consultant to perform Consultant's contract obligations.

4. **Staff.** Consultant is an independent Contractor and Consultant is not employed by Client. Consultant is hereby contracting with Client for the services described in this Agreement and Consultant reserves the right to determine the method, manner, and means by which the services will be performed. Consultant is not required to perform the services during a fixed hourly or daily time. Consultant shall not be required to devote his

full time to the performance of the services required hereunder, and it is acknowledged that Consultant has other clients and offers services to the general public. The order or sequence in which the work is to be performed shall be under the control of the Consultant. Client shall not provide any insurance coverage of any kind for the Consultant, and Client will not withhold any amount that would normally be withheld from an employee's pay.

5. **Non-Solicitation of Employees.** During the term of this agreement and for a period of twelve (12) months thereafter, Client agrees not to solicit, recruit, or employ any employee of Consultant without the prior written consent of the President or Chief Executive Officer of Consultant. Consultant hereby agrees that it will not solicit, hire, or retain, in any capacity whatsoever any of Client's employees without written consent from Client.

6. **Complete Agreement.** This Agreement contains the entire Agreement between the parties hereto with respect to the matters covered herein. No other Agreements, representations, warranties, or other matters, oral or written, purportedly agreed to or represented by or on behalf of Consultant by any of its employees or agents, or contained in any sales material or brochures, shall be deemed to bind the parties hereto with respect to the subject matter hereof. Client acknowledges that it is entering into this Agreement solely on the basis of the representation contained herein.

7. **Scope of Agreement.** The enforcement of this agreement shall be governed by the laws of the State of California. If the scope of any of the provisions of the Agreement is too broad in any respect whatsoever to permit enforcement to its full extent, then such provisions shall be enforced to the maximum extent permitted by law, and the parties hereto consent and agree that such scope may be judicially modified accordingly and that the whole of such provisions of this Agreement shall not thereby fail, but that the scope of such provisions shall be curtailed only to the extent necessary to conform to law.

8. **Additional Work.** After receipt of an order that adds to the Services, Consultant may take reasonable action and expend reasonable amounts of time and money based on such order. Client agrees to pay Consultant for such action and expenditure as set forth in this Agreement.

9. **Notices.**
(i) Notices to Consultant should be sent to:

Karl W. Palachuk
Small Biz Thoughts

_____

Sacramento, CA _____

(ii) Notices to Client should be sent to:

_____

_____

_____

_____

You should have a lawyer review every agreement or contract you sign.

10. **Disputes**. Any disputes that arise between the parties with respect to the performance of this contract shall be submitted to binding arbitration by the American Arbitration Association, to be determined and resolved by said association under its rules and procedures in effect at the time of submission and the parties hereby agree to share equally in the costs of said arbitration.

The final arbitration decision shall be enforceable through the courts of the State of California. In the event that this arbitration provision is held unenforceable by any court of competent jurisdiction, then this contract shall be binding and enforceable as if this section were not a part hereof.

LIMITED WARRANTY

11. **Liability**. Consultant warrants to Client that the material, analysis, data, programs, and services to be delivered or rendered hereunder, will be of the kind and quality designated and will be performed by qualified personnel. Special requirements for format or standards to be followed shall be attached as an additional Exhibit and executed by both Client and Consultant.

Consultant offers no guarantees or warranties, express or implied, as to system availability and functionality during any phase of its support services and makes no guarantees or warranties, expressed or implied, regarding the ability to resolve computer-related problems.

Consultant makes no other warranties, whether written, oral or implied, including without limitation warranty of fitness for purpose of merchantability. In no event shall Consultant be liable for special or consequential damages, either in contract or tort, whether or not the possibility of such damages has been disclosed to Consultant in advance or could have been reasonably foreseen by Consultant, and in the event this limitation of damages is held unenforceable then the parties agree that by reason of the difficulty in foreseeing possible damages all liability to Client shall be limited to the lesser of Ten Thousand dollars ($10,000.00) or the total amount paid labor to Consultant in the previous three whole months as liquidated damages and not as a penalty.

12. **Scope of Services.** Consultant will provide the following Services under this agreement:

Item 1. Nature of Service.
Consultant agrees to provide consulting services related to computer hardware, software, network configuration, Windows operating systems and networks, database development, and programming. Such work will be done on behalf of Client and will be performed at a location or locations to be determined by Consultant. Work may be performed at the client's site or remotely, at Consultant's discretion.

Item 2. Regular Consulting Hours.
Regular Consulting Hours shall consist of any time Consultant works for Client during regular business hours. Regular

business hours are defined as 8:00 AM to 5:00 PM Pacific Time Monday through Friday excluding national holidays.

Item 3. Extended Consulting Hours or "After Hours" Consulting Hours
Extended Consulting Hours or "After Hours" Consulting Hours shall consist of any hours worked by Consultant outside of the period defined as "Regular" Consulting Hours. This includes weekends, the period 5:00 PM to 8:00 AM during weekdays, and all holidays.

Item 4. Short-Notice Emergency Service
Short-Notice Emergency Service shall consist of any labor requested by the client that escalates an issue to "Priority One" outside the scope of Consultant's standard operating procedures. Whenever an issue is artificially raised to Priority One, the Service Manager will inform the client and designate the issue to be Short-Notice Emergency Service.

13. **Services and Responsibilities**. Support agreement calls are prioritized by severity, and then by the order in which each call is received. Critical equipment outages are treated with the highest priority. Consultant will make a good faith attempt to return every support service call within two (2) business hours from receipt of call, during normal business hours.

Client is responsible for maintaining a working broadband connection and agrees to cooperate with Consultant and Consultant's employees for the purpose of providing remote support and troubleshooting.

This Agreement is for services only. Client is responsible for the cost of all hardware, software, consumables, and related costs for repair or replacement of hardware not covered under warranty;

After-hours service is available as part of this agreement by a voice mail system that will notify a consultant's "on-call" technician of any emergency technical call that is received during off-hours, including weekends and holidays. Consultant's technician will make a good faith effort to assist with client emergencies by telephone or remote support tools during off-hours.

Service calls that cannot be completed during regular consulting hours will be completed by consultant during the next available regular consulting hours time slot. In the event that client requests that consultant continue to work after regular consulting hours, client shall pay the rate for Extended Consulting Hours.

14. Cost of Services. All General Consulting Hours, as defined in Section 12 "Scope of Services" will be provided at a rate of <u>one hundred sixty dollars ($160.00)</u> per hour.

All Extended Consulting Hours or "After Hours" labor, as defined in Section 12 "Scope of Services" will be provided at a rate of <u>three hundred twenty dollars ($320.00)</u> per hour.

All Short-Notice Emergency Service labor, as defined in Section 12 "Scope of Services" will be provided at a rate of <u>three hundred dollars ($300.00)</u> per hour.

These rates are subject to change, subject to a thirty (30) day written notice.

15. **Credit Terms.**

a) Managed Services fees (the flat monthly fees) are due and payable on the first day of each calendar month whether or not client has received an invoice from Consultant.

b) All invoices to Client for labor shall be due within 20 days.

c) All invoices for Hardware, Software, and other materials are to be paid in advance or "due upon receipt."

d) Any unpaid sums over 30 days old that are not in dispute shall bear interest at the rate of 1.5 percent per month. Costs of collection including reasonable attorney's fees shall be borne by the Client.

e) All late payments are subject to a twenty-five dollar ($25.00) late fee.

f) Consultant reserves the right to refuse service of any kind on all accounts with past due balances.

16. **Monitoring Software.** In order to provide the services specified in this Agreement, Consultant must install remote monitoring and management software on Client's servers, desktop computers, laptops, and possibly other equipment at Client's office. Client grants permission to Consultant to install

remote monitoring and management software from any remote monitoring and managing software company deemed necessary by Consultant.

17. **Services; Payment Terms.** Consultant agrees to perform for Client the following services (the "Managed Services") for the following monthly fixed fees. Check Appropriate Service (Platinum, Gold, or Silver).

a) **Platinum Support** as defined in the attached marketing material and updated periodically by Consultant.

19 Users, including servers, workstations, network equipment ($ 125 / User)
Monthly Total:_____$2,375.00

b) **Gold Support** as defined in the attached marketing material and updated periodically by Consultant.
_____ Users, including servers, workstations ($ 100 / User)
Monthly Total:_____$_____

c) **Silver Support** as defined in the attached marketing material and updated periodically by Consultant.
_____ Users, includes monitoring only. ($ 29.95 / User)
Monthly Total:_____$_____

In addition to the monthly fees set forth above, Client agrees to pay an initial setup of remote monitoring services ("Setup Services") fee in the amount of $__00.00__. This fee is 100% of the monthly recurring service charge.

Client shall pay the Setup Services fee (if any) upon execution of this Agreement.

18. **Determining the Number of Users.** The number of users is determined by the number of individuals in the Client's business who will have any machines or devices supported under this Agreement. The price is determined by the number of users and not the number of devices they use.

19. **Services Included in Managed Services.**

a) "Network Equipment Support" (part of the Platinum Package) shall consist of all labor related to maintaining configuration, logging (if possible and appropriate), and monitoring of network equipment, including routers, firewalls, switches, spam filters, and other equipment used to move, monitor, or intentionally affect Ethernet traffic on Client's local area network. Network Equipment Support shall also consist of working with Client's Internet service provider to maintain proper configuration of Internet equipment at Client's office, whether owned by Client or Client's ISP.

b) "Server Support" shall consist of all labor related to maintaining Client's server operating system, any programs included in the operating system, backup software, virus scanning software, hard disk defragmentation software, and the following programs installed after the operating system:

_____

_____

Client agrees that Client will maintain separate service agreements with these software vendors. Consultant will coordinate or provide all service related to these products. Consultant shall also provide all service related to SQL, Exchange, or SBS, or any other software installed on each server.

c) "Desktop or Laptop Computer Support" shall consist of all labor related to maintaining the computer operating system, any programs included in the operating system, Microsoft Office products, virus scanning software, and the following programs installed after the operating system:

_____

_____

Client agrees that Client will maintain separate service agreements with these software vendors. Consultant will provide all service related to these products.

d) "Mobile Device Management" shall consist of all labor related to monitoring, configuring, and troubleshooting mobiles such as tablet PCs, cell phones, and iPads. This Agreement only covers basic operational maintenance as well as monitoring if our agents can be installed.

Client agrees that Client will maintain separate service agreements with these software vendors. Consultant will provide all service related to these products.

e) Hardware Support: In addition to the

maintenance of the operating system and software, above, Consultant agrees to provide hardware support for all equipment that is purchased from Consultant and covered under this Agreement, provided that such equipment is less than three years old and is under manufacturer's warranty, or is over three years old and covered by a manufacturer's extended warranty.

20. **Software Updates.** Maintaining the systems described above shall include applying all appropriate software and operating system updates in a reasonable amount of time. Consultant shall determine when software updates are appropriate and what constitutes a reasonable amount of time.

Client acknowledges that if Client requests updates that Consultant considers inappropriate, or wishes to have updates applied before Consultant deems them safe, Consultant is not responsible for the consequences of such actions and Client may be charged a Regular Consulting Hours or Extended Consulting Hours charge, as the case may be, for all labor related to the consequences of such actions.

Furthermore, if Client performs or allows anyone other than Consultant to perform any maintenance on any of these machines, Consultant is not responsible for the consequences of such actions and Client may be charged a Regular Consulting Hours or Extended Consulting Hours charge, as the case may be, for all labor related to the consequences of such actions.

21. **Warranties.** All equipment (network equipment, servers, printers, desktop computers, laptop computer, etc.) must be under an original manufacturer's warranty, or some other similar warranty or extended service plan in order to be covered by this Agreement. Work performed on equipment that is not under warranty will not be covered under this Agreement. All such work will be billed according to the rates and terms of the Agreement. All software packages other than Microsoft software must be under a maintenance agreement in order to be covered by this Agreement. Work performed on software packages that are not under a maintenance agreement will not be covered under this Agreement. All such work will be billed according to the rates and terms of the Agreement.

IN WITNESS WHEREOF, the Parties hereto have signed this Agreement and agree that is shall be binding upon the parties and their respective heirs, successors, and assigns.

Consultant: _____
Name:_____
Title:_____

Client:_____
Name:_____
Title:_____

Section III
Getting Help and
Managing Agreements

Chapter 8
Getting Help

Chapter 9
Managing Your Service Agreements

Chapter 10
Conclusion

"The man who is too old to learn was probably always too old to learn."
— Henry S. Haskins

Chapter Eight – Getting Help

Defining your business, and your relationships with clients is important work. That's why it's also important to get help. In this chapter we discuss three types of help you need with regard to service agreements: Legal, Financial, and Insurance.

Most of us are not lawyers, so we don't spend a lot of time poring over contracts and agreements of various kinds. While we can usually make sense of these things, it's not really where we shine.

The same is true with finances. If you're good at business, you're probably also good at "running the numbers" or penciling out whether a project is profitable and worthwhile. But that's different from knowing the tax ramifications of defining different types of relationships.

For both of these areas, we are wise to rely on professionals who can help us make the right decisions. Then there's the question of insurance. You may or may not decide to get Errors and Omissions insurance. At a minimum, it is worth considering.

Legal Advice

There are lots of reasons to engage an attorney for your business. I use one law firm for intellectual property (trademarks, copyright), one for contract reviews, and a third one for employee matters.

You may find one attorney or law firm that does everything you need. You may also decide that you don't really need an attorney and that you can just piece together some contract language you've come across. You would be wrong.

Please don't be penny-wise and pound-foolish. When it comes to service agreements:

You need an attorney.

Let me step back a bit and give a little theory about lawyers and such. In Chapter Four on "Credit Agreements" we discussed why you need *something* rather than having nothing to define your relationship with the client.

If you buy the argument that you need something, then as the relationship becomes more formal, and the stakes become higher, you will see that the need to protect your interests becomes greater.

Here's what an attorney will be able to do for you:

- Verify that the document makes sense and is complete
- Verify that the document fulfills your goals
- Check to make sure there are no conflicting elements within the document
- Make sure that the document is enforceable in a courtroom

Some people get discouraged when they give a document to a lawyer and get it back with just a few changes and a bill for $250. But think about your own consulting business. Let's say someone is afraid that they have a virus and they pay you to check out their PC. You run a thorough scan and examine the behavior that made them bring it up.

If you don't find a virus, have you still provided a service? Yes. Should you be paid? Absolutely.

You should have a lawyer review every agreement or contract you sign.
www.smallbizthoughts.com

There is a tendency for lawyers to change things. After all, you're paying for something. I believe you can take a document from one attorney, go across the hall to another attorney, and the second guy will make changes. It happens.

Sometimes these changes are stylistic or personal preference. Sometimes, they affect the ability to keep that part of the agreement from being thrown out of court.

If you're not sure why a change was made, ask. After all, you're paying the bill.

How Often Do You Need an Attorney?

I try to have an attorney review my general agreements every few years. After all, laws change all the time.

In our business, we have some "standard" agreements. When we sign a new client, we switch out the client info and print the new agreement. With such a change we feel very comfortable that we haven't invalidated anything.

But sometimes a client will make a good suggestion, or require that something new be thrown into the agreement. A perfect example of this is when we got a new client who required that we be a "Drug Free Workplace."

Sometimes we run these changes by the attorney and sometimes we don't. But after a few years we realize that we changed a little bit here and a little bit there. Pretty soon we have a Frankenstein contract, patched together with pieces of other contracts.

At that point, we call in the professionals to give Frankie a facelift.

Write Your Own Draft

Having said all that, I'm sure you're 100% onboard with the idea that you need an attorney. But that doesn't mean you need to personally put their children through college.

You can save yourself a lot of money by drafting your own service agreements (and all other legal agreements) and asking the lawyer to review them. There are several reasons this makes good business sense as well as financial sense.

Unless you happen to stumble upon a lawyer who specializes in service agreements for technology firms, you know more than the lawyer about what you need. If you went to her and said "Draft a technology services agreement," your lawyer would spend a lot of time floundering around looking for examples as a starting place, followed by personalizing them for you. All billable.

If you take a draft of what you want to your lawyer, she'll probably glance at it and quote you a price for dotting the i's and crossing the t's.

In general, whenever you want to have legal work done, it is a good idea to write the first draft yourself. You know what you want. Sometimes it is hard to put it in words. Ask for help from friends, family, business associates, etc.

When I was teaching college, I used to tell my students to just start writing something because you can't edit a blank page. Start with an outline. What do you need? What does the client need? Whose tail are you protecting? Draft something and your lawyer will have a place to start.

How Are Attorneys Paid?

Aside from the obvious answer ("Attorneys are paid very well, thank you."), there are several ways to pay your attorney. After all, their business is not dramatically different from yours. You're both in the service business and you both provide labor in exchange for a fee.

Most attorneys will agree to an initial meeting to discuss the services they offer, the prices they charge, and to see whether they're a "good fit" for your project. Most will not charge for this initial consultation.

I don't know why I need to say this, but I need to say this: Don't be afraid to discuss fees with your attorney at any time. Ask what's billable and what's not. Ask "Is this on the clock?" Remember, we're talking about your money!

Easily ninety percent of all attorneys are happy to discuss money. Don't hire the other ten percent.

The two most common billing arrangements are by hourly fee and by flat rate. See? I told you this would look familiar. Our trademark attorney charges by the hour. Period. Every time I call and he picks up the phone, I get a bill in quarter-hour increments.

The attorney that works with our employee-related matters prefers to quote a flat rate per job. When I asked him about creating the drug-free workplace addendum and forms for employees to sign, he did a little research to see what's involved and then quoted me a price. I don't know whether it took him ten minutes or three hours. The price was fair, the work was done in a timely manner, and I was happy to pay it.

Finally, we have an attorney who reviews our contracts and service agreements. The first go-through she charged by the hour. Now, I ask what she'd charge and she quotes a flat price.

All three attorneys know that I pay my bills and they don't mind just doing the work when I call. If I ever get a bill I don't expect, I call and we chat. Just as you would expect from your own clients.

The other major way that attorneys are paid is on a contingency. This does not apply to you. Contingency means that the attorney doesn't get any money now, but will take a slice of the pie if you win a lawsuit. As you can see, this only applies if you're suing someone, which is way outside the scope of this book.

The only other variable with lawyers is whether they want a retainer. Some attorneys ask for some money "up front" before they proceed. You probably won't see this very often on small projects.

If you're doing a trademark search, which is a long process with a well-known timeframe, you might be asked to put up a retainer as well as to make payments at specific stages of the process. For work related to contracts and service agreements, you're not likely to be asked for a retainer.

Working With Attorneys: Two Simple Rules

There are two little "odd" behaviors I see with consultants and their attorneys. The first is a tendency to ignore advice. The second is to expect simple answers.

Did you ever have a client who asked you a very specific question, listened carefully, and then did something else? If you're like me, your response was "Why does this guy ask my advice—pay for my advice— and then not take it?"

It's the same way with attorneys, except that you're the client. You pay good money for good advice. So simple rule number one is:

Take Your Attorney's Advice!

What is it about people that they'll take advice from a stranger at the hardware store, or a sixteen-year-old at the home electronics store, but don't want to take the advice of a professional after paying hundreds of dollars for that advice? Yes, you know your business better than he does. Yes, you've been doing this a long time.

But you don't know the law and you don't understand the ramifications of all the decisions you make. When you walk into a new client's office, you can quickly know more about their computers than they do. An attorney is the same way with your business. A quick review of your contracts will tell him more about the legal side of your business than you know yourself.

To get the greatest value from your attorney — as with your financial advisor — you need to treat this person as a member of your team. It's still your business and you still get to make all the decisions. But please take the advice you're paying for.
If you think there's a legitimate reason to not take your attorney's advice, seek another opinion. Perhaps you need to find someone whose approach to business is more consistent with your own. But if you get the same advice twice, you really need to take it.

Sometimes, when we work with attorneys, we are under stress or in a hurry. In either case, we want to spend as little money as possible and get an answer as quickly as possible.

This leads to simple rule number two:

Don't expect your attorney to give a simple answer to a complex question.

Sometimes life, and business, get complicated. Trying to create a nice little summary of a problem and get a simple answer is natural. Unfortunately, it doesn't always make sense.

At the end of the process, when you've spent adequate time and gotten adequate advice, you'll be glad that you spent the time. But you have to go through the process, even when you're in a hurry.

It's always better to do something right the first time than to do it over. If nothing else, it will cost you less money in the long run.

Consider another example from your own business. Sometimes a client will come to you with a request that doesn't seem to make sense. For example, a client might say "I want you to disable virus scanning on the file server." Do you do it? Probably not.

I hope your response is to ask "Why do you want to do that?" At this point, the client will explain that files are very slow to open, that he did some research on the Internet, and this seems to be the answer. Your job as a trusted advisor is to guide him to a more systematic and effective approach.

In this case, the client came to you with a request for a specific action to take, but he should have come to you with a statement of the problem and asked you for your advice. After all, without doing systematic troubleshooting into server message blocks, patch levels, etc. it would be irresponsible for you to simply fulfill the request that was originally made.

Attorneys have the same problem. You might come in and ask for a paragraph that says clients can't apply fixes on their machines if they have a contract. That sounds simple to you, but it may not be simple to the attorney whose job is to write an agreement that's in your longer-term best interest.

How Many Contracts Do You Need?

We try to maintain one standard Credit Agreement, one standard Hourly Agreement, and one standard Flat-Fee Agreement. We sell remote monitoring as an item within the flat-rate agreement.

If you sell software development services, outsource your I.T. people, or work with other special projects, you may have additional agreements.

You should have a lawyer review every agreement or contract you sign.
www.smallbizthoughts.com

As a general rule, you should have as few different agreements as you can get away with and as many as you need to cover your assets.

Financial Advice and Insurance Coverage

The other two types of "help" you need are Financial and Insurance.

There are many options in both of these categories. We won't go into a complete breakdown of all your options. The focus here is on supporting your service agreements.

For the financial arena, you have many options. Whether it's an accountant or an enrolled agent, you need to find a good financial mind to make sure you're setting things up properly.

Two Definitions

A *Certified Public Accountant* is an accountant who has passed a battery of exams to demonstrate proficiency with taxes and finances. CPAs must also fulfill ongoing educational requirements and pass state-level requirements. Not all accountants are CPAs.

CPAs are allowed to "bless" financial statements by including an opinion letter regarding the reliability of the financial statements. In such cases, the CPA is liable for any errors that he misses. As a result, CPAs are allowed to charge more than a non-CPA for their services.

An *Enrolled Agent* is a taxpayer's advocate, authorized by the U.S. Department of the Treasury to represent taxpayers in IRS audits. They are experts in income tax preparation as well as all aspects of tax audits, collections, and appeals. In addition to passing exams, Enrolled Agents also take continuing education classes to keep up to spec on tax code changes.

If you have your contracts reviewed by a tax attorney, that might include all the financial advice you need. Just make sure that your tax attorney also has the experience you need with regard to contract reviews.

So, what are you looking for financially? There are two main reasons for keeping track of finances. One is to make reports to third parties (most often the government, and most often associated with taxes). The other is to help manage your business.

In terms of a financial advisor, you need someone to help you with the decisions from Chapters One and Two. How you define your business (sole proprietor, corporation, etc.) can have a dramatic effect on taxes and how your money flows.

That's why we're only discussing tax-focused financial advisors, such as CPAs or Enrolled Agents. The more you can lay out "the big picture" about how you want to structure your business, the more benefit you'll get from your financial advisor.

The very decision about whether to incorporate, for example, could be based on finances without regard to the question of personal liability. So, while limiting liability is a major benefit, it might not be a consideration for you.

You need someone who will know tax rates and special rules for various forms of business in your state. In addition, specific types of labor sales or services might be taxed at different rates.

And don't forget the ongoing costs. If you're going to spend an additional $1,000 per year in tax preparation with one decision over another, that may be a significant factor.

Every once in a while, there's a major overhaul of the tax code. The U.S. is going through this as I put the final touches on this chapter. No one from either party knows wat the final, final, final tax code is going to look like. Even after it passes and becomes law, it might be a year before the tax consequences are clear to tax professionals!

So, should you change from an LLC to a C-Corp? Would your taxes go up or down? I don't know. No one knows until they go through your records and create a realistic tax estimate.

Oh . . . and those new tax codes will be changed and updated every year for the rest of your life. Maybe not big changes every year, but some changes every year. Ignoring all these changes and not seeking advice can result in paying a lot more in taxes than you need to!

Making Sure You Have the Right Financial Pro

You may already have a great tax preparer. But that may not be exactly what you're looking for. You want someone who can easily discuss all those things we covered in Chapter One.

Ideally, you'll find someone who can grow with you and who wants to do more than just tax returns. Don't get me wrong, tax returns are important. But you also need an advisor.

You need someone who can help you discuss the legal form of your business from a financial perspective. You need someone who can give you a ballpark of your tax situation at least once or twice between tax returns.

You need someone who will meet to discuss strategies for new pricing options, and who will help you "run the numbers." For example, if you're considering a flat-rate component, how does that need to be structured to maintain profitability?

Finding this financial guru is similar to finding a good attorney, with one exception. Many people already work with a financial professional of some type. If you do, you have the difficult decision to make about whether that person is the right person for the new responsibilities you require.

As I mentioned earlier, we tend to become personally attached to our financial advisors. That makes it difficult to move on when you need more than your advisor is able to give.

Just remember that your loyalty has to be to your business. You can't go forward with someone you know is wrong for the job. That's just like keeping an employee you know should be let go.

If it's just too emotionally painful to switch advisors, then you need to re-evaluate where you want to go with your business and make sure your plans fit within the comfort zone of your financial advisor. That plan doesn't guarantee that your business will be successful. It just guarantees that your advisor is comfortable.

I don't mean to put too sharp a point on it, but this is an extremely important decision for your business. You need to make sure you make the decision and don't just go with the status quo because it's the easiest thing to do.

If you decide to find a new financial advisor, getting referrals is always the best place to start.

Finding the Right Help

How do you find a good lawyer or financial advisor? Luckily, most I.T. Consultants have at least one of each as a client. You might not be comfortable using a client as your attorney or tax pro, but they're still a good resource. Ask them for a referral!

You might also ask clients or other consultants who they use. Remember that you're looking for someone who does this specific type of work, and it would be great if they had some experience in the technical arena.

As a general rule, the phone book is not a good place to find an attorney or an accountant. There are hundreds of ads, they're not going to list exactly what you're looking for, and when you call them, they'll all say they do it.

You're looking for someone who's as good at her job as you are at yours.

Side Note: You should always ask your clients for referrals when you're looking for something. They're able to pass on leads and this helps them in their other business relationships.

And most importantly, find someone now. Don't wait until you need someone. Like everything else, this process takes time.

Luckily, there are many online search tools for finding attorneys, accountants, and Enrolled Agents. In addition to your local bar, accounting, and Enrolled Agents association, check out these sites:

- LexisNexis www.lawyers.com
- The National Bar Association www.nationalbar.org
- The American Bar Association www.abanet.org
- Your State Bar Association
- National Association of Enrolled Agents www.naea.org
- State-level Enrolled Agents associations
- American Institute of Certified Public Accountants www. aicpa.org
- CPA Directory www.cpadirectory.com
- State-level CPA associations

Insurance Coverage – Errors and Omissions

Finally, we come to Insurance, Errors and Omissions.

Some clients require E&O Insurance. If that's the case, your decision is made. If not, you have to decide whether you need it. There is legitimate debate about whether you need E&O insurance. After all, very few I.T. Consultants ever get sued.

If you're good at what you do, and you've been in business for very long, you've come across a scenario similar to this:

> Dewey Cheatum & Howe, Attorneys at Law, fire their computer consultant because he accidentally deleted the data directory and their backup doesn't work. They call you in. The former consultant won't give them the password to their server or the firewall.

> You go to reinstall a program on the server and discover that he never delivered any of their software. It might be legal, but there are no DVDs or licenses of any kind for any software in the office.

> An examination of the hardware reveals that the lawyers purchased an old server they thought was new. It's still registered to the old techie.

And so forth. And so on.

At no point is there any discussion about suing this clearly-incompetent consultant who appears to have left them running their business on illegal software.

I've seen situations like this several times. I use the attorney example because the two worst cases I know about involved lawyers. And if they're not going to sue, who is?

My point is this – the chances that you'll be sued are extraordinarily slim. And if you're competent, they're even slimmer.

Having said that, please review the discussion in Chapter One. Depending on how you're organized, you may have your personal finances hanging out there waiting to be lost in a lawsuit. How many lawsuits can you survive? Having signed agreements — especially with arbitration clauses — helps a lot. A properly maintained corporation protects your personal assets. That helps, too.

But at some point, the cost of insurance will be small compared to the chances of losing everything. Here's a fact – if your business is successful, you will eventually buy E&O insurance. After all, a successful business is an asset unto itself and that asset needs to be protected.

The only remaining question is whether you'll buy E&O insurance sooner or later.

Just so you know the point of pain, through some Internet chats, a little research, and personal experience, E&O insurance can be had in the range of $1,500 to $2,000 per year. In some states it might be closer to $500. So shop around.

Even though I think my chances of being sued are very low, my company has E&O insurance. As long as you're chatting with a lawyer and a tax advisor about your service agreements, it's worth asking them what they think about it.

If your profit is low, you don't own a house, or your corporation has no assets (and you've rigorously protected the corporate veil) then maybe $1,500 per year is too much. But when your profit is what you consider "high," you do own a house, or your corporation has assets that could be lost in a lawsuit, then you'll pay the money.

Callout: What is E&O Insurance?

Errors and Omissions (E&O) Insurance covers mistakes you make—things you do and don't do. For example, if you write code that causes a problem with a database, that's an error. Whether you did the wrong thing or didn't do the right thing, you're covered.

Similarly, if you present to the client that their backup is working great, but something bad happens and there's no backup, you would rely on your E&O Insurance to protect your assets.

Disaster Recovery activities are often excluded from E&O policies. If you manage disaster recovery for your clients, you'll want to look into policies or riders that will cover you in this area.

E&O is not Liability Insurance

E&O covers your job (doing it right). Liability insurance covers damages that result from actions taken by you or your employees. E&O covers consulting-related mistakes.

For example, if you drop a computer on someone and break that person, you may be liable for medical bills. Or if your employee gets mad and takes a hammer to a monitor. Or if you move a shelf and knock an antique vase onto the floor.

Stuff happens.

In general, liability insurance covers you for all the things that can go wrong as a result of something your company does. Prices vary, of course, but liability insurance starts in the range of $1,000 per year.

E&O is not Property Insurance

If you keep very low, or no inventory, and if you have a home office, then your homeowner's policy may cover the loss of computer equipment due to theft, fire, etc. Check with your current insurer. Your homeowner's policy may also cover contents of your car. Ask.

If, however, you have an office or storage facility, and find yourself with a large dollar amount of equipment on hand, then you should consider property insurance.

After ten years in business, we decided to get property insurance after one of our clients had every piece of computer equipment stolen. We ordered up the new equipment for him. At that time our office had a new server, a new color laser printer, a new workstation, and some older workstations.

When we added all the replacement equipment for our client, I started to get nervous. We also had another client who had ordered a server, so it was in our office along with one or their workstations.

I looked around the office and realized that our exposure was higher than I was comfortable with. Since I was very aware of the fact that one could have every piece of equipment stolen at once, I got pretty nervous.

Rather than sleep in the office, I signed up for property insurance. Again, starting price is about $1,000. You may need to shop around if you want to find property insurance without liability insurance.

Insurance Audit

At least once every two years you should do an audit of your insurance coverage. This is particularly true if you are growing or changing the way you do business. If you add products (such as managed services), your E&O insurance may go up. But if you sign contracts with 100% of your clients, your insurance rates will probably go down.

Think about where all the pieces of your business are. If you have a server in your house, verify that it's covered by your homeowner's insurance. As long as you have them on the phone, ask about equipment in your garage and in your car. What about equipment in your employees' cars?

Homeowners' insurance won't cover equipment at your office. Do you have a policy that does? What is covered by your errors and omissions policy? Are there any important pieces of your business that are not covered? Also find out about liability insurance in your home, at your office and, most importantly, at your clients' offices.

Here are a few resources for insurance. This is just the tip of the iceberg, and not an exhaustive list by any means. Please do an Internet search or ask other consultants about what works for them.

- The ASCII Group www.ascii.com
- IT Insure www.itinsure.com
- Sadler & Company, Inc. www.insurancefortechs.com
- Techinsurance www.techinsurance.com
- Tech Shield www.techshield.com

Concluding Thoughts

Insurance seems far afield from service agreements, but it's part of the structure you need in place to run a managed service business. With rare exceptions, the only kind of insurance you're likely to have a client insist on is E&O. But you should take some time to look into the others, especially as your business grows.

As for lawyers and financial advisors, I recommend that you find them before you need them. Give due consideration to the amount of money at stake and weigh this against the cost. Consider two factors.

First, consider how you want your clients to think about your services. Are you "too" expensive and not worth the money? No. Of course not. The work you do actually saves your clients money down the road. In addition, when you do the work, you know it's done right. You're not a dentist or a plumber trying to be a computer professional.

In the same way, you should recognize that professionals in law and tax/ finance are the right people to call on for you. Don't be the computer geek who tries to do the work of a CPA or an attorney.

The second thing to consider is the very valuable role these professionals can play as your advisors. A true advisor knows about your business, knows something about your philosophy and approach, and can give you excellent advice that's tailored to your business.

For example, any tax pro can tell you the criteria for taking a certain write-off. But an advisor can look to your past and know enough about your future to give advice on whether this tax write-off is good for you, given the specifics of your business.

We've all seen clients who start out unwilling to spend any money. After a few years they come to the realization that they need to spend money on technology and that it's actually good for their business.

You're the same way. If your business is to grow, you will eventually see the value of using lawyers and accountants and insurance to protect your business.

Callout: Working with Professionals

Q: When are people most likely to spend money for help?

A: When it saves them time or when they can't do the work themselves.

In the case of hiring help, we all make these calculations.

I can change my own oil. But with driving to buy the oil,

getting all set up, changing it, and disposing of it, I'll waste at least an hour of my life. Or, I can get it done by someone else for $3.95 along with a free inspection, and I don't have to wash my hands.

We all have a variety of skills somewhere on the continuum between "No Knowledge" and "Complete Competency" with regard to legal and financial matters. We all have to make calculations about how much knowledge we have and how much we need to buy.

I mentioned earlier that my parents did tax return preparation. Because of this, I have confidence that I can learn the rules, apply them, and do my own taxes. But, also because of this, I know that there's a lot of work involved in coming up to speed in this area. In addition, the rules change every year — sometimes quite dramatically.

As a result, I happily pay someone else to take care of my taxes. I already have a fulltime job! To me, it's a simple calculation.

Here's another example from our industry. A few years ago, I bought into the Robin Robins' Technology Marketing Toolkit (www.technologymarketingtoolkit.com). I paid a lot of money for this, but I made a quick calculation.

Just as with tax preparation, I don't think there's anything in the Technology Marketing Toolkit that a smart person couldn't education himself on. But, just like tax prep, I already have a fulltime job. As a result, I'm willing to buy the knowledge of a marketing strategy that's proven to work in my industry.

If you have a partner who's an attorney or a tax professional, great. But if not, you need to make an honest calculation about what it really costs to buy the knowledge to do something the right way versus trying to do it yourself.

"I've always felt that if you could develop an answer to a need, this was the way to make money. Most people are more anxious to make money than they are to find a need. And without the need, you're working uphill."
— Bill Lear

Chapter Nine – Managing Your Service Agreements

So now you have a collection of service agreements. As with anything else that's important to your business, this joyful new addition comes with some added work. Once you have agreements you'll need to keep track of them, make sure they're renewed on time, and make sure you use them to your company's best advantage.

This chapter covers a series of administrative chores. We discuss getting agreements signed, keeping your agreements organized, and finding tools to help you actually perform the "managed services." We finish with some sample cover letters to send out with your agreements.

Execution

The first step in organizing your contracts is to decide how they will be executed (signed). Many clients will not have e-signature programs, but you absolutely should. These are widely available from Adobe, Docusign, and other services. If you don't use one of these services, you should get a "wet" signature or exchange scanned document via encrypted email.

My preference is to use Docusign (www.docusign.com). My second preference is to execute via pdf, signed, and emailed as encrypted file.

Electronic signatures are so easy that clients generally execute right away. I had a coaching client who had never signed contracts with clients. I told him to just send them out and see how many came back – then focus his energy on those who actually delayed. Virtually every client just signed the document digitally and quickly.

If the agreement is a renewal, your cover letter should state whether changes have been made in some key areas. If no changes are made, you should say that as well. If no changes are made, the client is not likely to delay even a moment in signing.

The key elements that might change are: rate for regular hours, rate for extraordinary hours, and minimum hours to be charged for onsite and remote work.

If a client has asked for a change (e.g., you need to comply with an industry-specific policy), you should note that you have added a section to address their request.

Sales Tips for Service Agreements

Whenever we quote any job that we think will fill ten or more hours, we mention the service agreement option. We often get an opportunity to discuss this on an initial visit or sales call.

Whether the client expresses any interest or not, we quote the work based on a signed service agreement and we include two copies of the "proposed" agreement, executed and ready to sign and return.

We make a note on the quotation that these are the special reduced rates that assume a signed agreement. Of course we also note that our standard rate is \$_____. In our case, the contract rate is a nice clean 10% lower than the standard rate.

As with any selling, the easier you make it to say "yes" the more sales you'll make.

Staying Organized

When you sign your first agreement, all of your attention turns to delivering what you promised. For more information on that, see the next section. But you also need to put some energy into creating a business that manages service agreements. After all, the goal is to sign a second, and a third, and a fourth agreement.

If your agreements expire in, say, one year, then you don't really have to worry too much about managing your agreements for a year. But, if you're successful, you'll have twenty-five agreements when the year is up. And they'll all expire at different times. Some will require rate increases.

Note: There is an obvious benefit to a service agreement that automatically renews every 12 months! For the last fifteen years I owned managed service companies, I only signed service agreements that renew every month.

If you don't keep all this straight, you'll lose money. Therefore, it's in your best interest to put some kind of system in place to keep track of these agreements. A bit later we'll discuss a "real" PSA (Professional Services Automation) system. For now, we'll start with a basic tool we all have available – an Excel spreadsheet.

The following sample spreadsheets are included on the downloadable content that comes with this book (register at www.smbbooks. com). The first one is used to keep track of clients who have only signed the one-page credit agreement and not a "contract" or service agreement. The second one is for keeping track of clients who have signed agreements.

You should have a lawyer review every agreement or contract you sign.
www.smallbizthoughts.com

The Month First Billed tells us how long these folks have been a client. While the whole process we're going through seems to be about money, there's a little something special about a client who has been relying on us for five or six years. If nothing else, an anniversary card might be in order!

Non-Contract Rates

Updated   6/6/2018

	Client	Last Billed	Month First Billed	Last Rate Increase	Tech Rate Regular	Tech Rate Extended	Rate Prog.	Min. Onsite
1	Zelda, Inc.	1/1/2018	Jan 16	NA	$150	$300		1.0 hrs.
2	Yolanda, Inc.	1/13/2018	Mar 05	Jan 15	$150	$300		1.0 hrs.
3	Xerxes, Inc.	9/22/2017	Feb 17	NA	$150	$300		1.0 hrs.
4	Westley, Inc.	12/31/2017	Jun 14	NA	$120	$240		1.0 hrs.
5	Velma, Inc.	3/9/2018	Feb 04	Jan 15	$150	$300		1.0 hrs.
6	Unity, Inc.	12/30/2017	Apr 15	NA	$150	$300		1.0 hrs.
7	Thomas, Inc.	6/30/2017	Jul 16	NA	$150	$300		1.0 hrs.
8	Stella, Inc.	12/28/2017	Dec 16	NA	$150	$300		1.0 hrs.
9	Raph, Inc.	2/21/2018	Mar 15	NA	$150	$300		1.0 hrs.
10	Quigley, Inc.	3/14/2018	Aug 15	NA	$150	$300		1.0 hrs.
11	Printers, Inc.	5/5/2018	May 06	NA	$120	$240		1.0 hrs.
12								

Contracts Schedule

Updated   5/5/2018

	Client	Mnthly Maint.	Moni-toring	Month First Billed	Current Contract Dates	Last Rate Increase	Tech Rate Regular	Tech Rate Extended	Rate Prog.	Min. Onsite	Min. Remote	Notes
1	ABC, Inc.	X	X	July 2013	7/2017 -	NA	$120	$180.00	$150	1.0 hrs.	.25 hrs	
2	Def, Inc.	X		April 1, 2015	4/2015 -	NA	$120	$180.00	$150	1.0 hrs.	.5 hrs	
3	GHI, Inc.			August 2014	12/2014 -	5-Dec	$120	$180.00	$150	4.0 hrs.	.5 hrs	
4	JKL, Inc.	X		October 1, 2013	9/2017 – 06/30/2016	NA	$150.00	$150	.5 hrs	.25 hrs		
5	MNO, Inc.	X		October 1, 2010	08/2017 - 06/31/2018	May-05	$85	$150.00	$150	1.0 hrs.	.25 hrs	5/2/6 Renewal mailed
6	PQR, Inc.	X	X	August 1, 2017	08/2018 - 02/28/2019	Mar-03	$85	$150	$150	1.0 hrs.	.25 hrs	
7	STU, Inc.	X	X	April 18, 2006	4/2016 -	NA	$100	$150.00	150	3.0 hrs.	.5 hrs	
8	VWX, Inc.	X	X	April 1, 2016	4/2016 -	Apr-06	$120	$180.00	$150	1.0 hrs.	.5 hrs	
9	YZ, Inc.			February 1, 2015	08/2017 - 5/31/2018	NA	$100	$150.00	150	1.0 hrs.	.25 hrs	5/2/6 Renewal mailed
10	ABC, LLC	X	X	April 1, 2014	4/2016 -	NA	$120	$180.00	150	1.0 hrs.	.5 hrs	
11	Def, LLC	X	X	August 1, 2017	4/2016 -	Apr-00	$120	$180.00	150	1.0 hrs.	.5 hrs	
12	GHI, LLC			September 1, 2005	9/2017 - 8/31/2018	NA	$100	$150.00	150	1.0 hrs.	.25 hrs	
13	JKL, LLC	X		January 1, 2012	12/2016 -	Dec-06	$120	$180.00	150	1.0 hrs.	.25 hrs	
14	MNO, LLC	X		May 1, 2016	08/2017 - 7/31/2018	NA	$100	$150.00	150	1.0 hrs.	.25 hrs	
15	PQR, LLC	X		June 1, 2011	5/2016 -	May 06	$85	$180.00	150	1.0 hrs.	.5 hrs	
16	STU, LLC	X	X	April 1, 2017	1/2016 -	Jan-06	$120	$180.00	160	1.0 hrs.	.25 hrs	
17	VWX, LLC			August 1, 2015	9/15/2017 - 8/31/2018	na	$100	$150.00	150	1.0 hrs.	.25 hrs	
18	YZ, LLC	X		July 1, 2014	12/2015 -	5-Dec	$120	$180.00	150	1.0 hrs.	.25 hrs	
19	ABC Company			November 1, 2005	11/2017 - 11/30/2018	NA	$100	$150.00	150	1.0 hrs.	.25 hrs	
20	Def Company	X		February 1, 2010	8/1/2017 - 5/31/2018	Feb-01	$85	$150	$150	1.0 hrs.	.25 hrs	5/2/6 Renewal mailed
21												

Sometimes we are reluctant to raise our rates for people who've been clients for a long time. In fact, these are the people who are least likely to raise a stink when they see a price increase. Look, for example, at the folks who have not seen a rate increase in three years but who are still getting a reduced rate.

One of the main purposes of this spreadsheet is to manage regular increases for these people who are not on a contract. If you've been raising rates and still have people at the old rates, you're making your life more difficult. Remember, these folks need to move to service agreements. Preferred treatment is to be found there, not here!

Note that we do not have a programming rate for clients without a service agreement because we don't offer this service to them. We also do not have a remote service minimum because we do not offer remote service to them. Membership has its privileges.

On the spreadsheet for Contracts, we see similar fields, but we also have the following fields: Client gets monthly maintenance, Client gets remote monitoring, and the contract end dates. You can see that some of these agreements have end dates and some do not. Clients without a contract end date have agreed to an agreement that renews automatically until cancelled.

The highlighted rows represent clients whose service agreements are going to expire and who need to sign a new agreement. I have an administrative assistant go through these every month and highlight the agreements that need attention. We then draft a cover letter. At the end of this chapter we have several sample cover letters for service agreement renewals.

In addition to giving you a quick glance of how your business is doing, these spreadsheets also give you a more objective view of your rate structure. We all find it difficult to raise rates, especially with clients who've been with us for a long time.

But when you look at these spreadsheets and see that Client A is on a signed agreement, paying $120 per hour while Client B is without an agreement and also pays $120 per hour, it just doesn't look as fair. This is particularly true since Client A, with the signed agreement, is probably buying many more hours per year than Client B.

While I don't post these numbers where everyone can see them, I do make them available to the Service Manager. After all, he has to have some idea what he's managing. If clients ask what it will cost for a given job, he can give them a ballpark estimate. These numbers are also available to the folks who do the billing, of course.

The result of having other people see this picture of billing practices at a glance is that they keep me honest and help me to make more money. Because they don't have the emotional connection I do to long-term clients, these folks look at the spreadsheets and say things like "Well, She's going to see an increase this year." I then have to justify why we would not bring that one client in line with our current rates.

Keeping track of these agreements can also be done with more sophisticated software, which pops up reminders and may even send you an email. We still like to keep it in a spreadsheet as well for the "quick glance" advantage.

Tools to Help Deliver Managed Services

Signing all these agreements, and having so many that you need spreadsheets to keep track of them, all sounds very exciting. But how do you deliver these services? Even with a team of people, how does one manage the actual delivery of managed services?

From my perspective, there are two options. You can buy software tools to help you manage your business and deliver services, or you can "roll your own" with a collection of free tools and applets. Most people start out rolling their own. If you support Microsoft Servers, this is not too difficult of a task.

If you haven't started rolling your own, you're probably wondering what's involved. Here are the basic chores that need to be accomplished:

1. Managing the Business Side

- Creating and tracking Service Requests
- Monitoring time spent on each job or task
- Managing employee time
- Invoicing clients
- Generating reports about profitability

2. Monitoring systems and Delivering Service

- Working Service Requests
- Monitoring Servers, desktops, laptops, and devices
- Applying Windows Updates, virus signature updates, program updates, service packs, etc.
- Preventive (monthly, weekly) maintenance
- Generating alerts when servers are down or other critical events occur
- Performing offsite backups
- Providing remote support to servers, desktops, and laptops

If you intend to roll your own, you can actually do a great job with the tools that come with Windows Servers, WSUS, and one "outsider" tool.

When we started offering Remote Monitoring as a separate service, we just connected with Terminal Services over a VPN. With SBS 2003, Microsoft introduced a secure replacement for this via Remote Web Workplace. In fact, it gave us much faster access to desktops.

All updates were pretty manual until about 2005 when WSUS (Windows Server Update Services) was released. WSUS has some great scripting options and is very powerful. It also has a steep learning curve, so we can't just hand it over to a new technician.

Better options include almost any of the professional RMM (Remote Monitoring and Management) tools, such as SolarWinds MSP, Auvik, and many others. In addition to patch management, all good RMM tools include or ship with a remote control agents and a preferred anti-virus tool.

Virus updates and monitoring backup logs have been easy jobs for a long time. All of the major anti-virus programs have a version that can be pushed from the server to the desktops. Some have better reporting than others. Similarly, the major backup programs offer remote email alerts as well as excellent logging of backup jobs.

Some people are happy with the NT backup program that ships with Windows Server. But today, most people prefer something more robust such as a BDR – Backup and Disaster Recovery tool. BDRs tend to give much faster recovery time, and an option to back up to cloud storage.

We never delved into offsite backups when we were rolling our own because we never found a reasonably priced package. We offer this service now because we got a great price along with our managed service software.

The only other tool we used for rolling our own was a program called Servers Alive. This great little program used to be free, and it's not too expensive if you're monitoring only a handful of servers.

Servers Alive (www.woodstone.nu/salive) will check basic connectivity via ping, pop3, smtp, http, https, snmp, DNS, and several other options. When a server or service fails, it will send you an alert and put the entry in a log. It's a great sales tool when you get a text message as a customer's server is rebooted as part of monthly maintenance.

Of course, on the business management side, most of us use QuickBooks or a similar "accounting" package to keep track of all the financial stuff. And when it comes to accounts receivable, invoicing, balancing checkbooks, and so forth, QuickBooks is still the king.

But for generating and keeping track of service requests, you really need something more specialized. Just as your clients have line of business applications for construction or legal or plumbing, you also need a line of business application for running you SMB Consulting practice.

Here again there are two options. You can buy a package built specifically for our type of business, or you can buy a more generic customer relationship management system and make it fit our type of business. The term PSA – Professional Services Automation or Administration – has become widely accepted as the term to describe LOBs built for technology consultants.

Both of these options take a great deal of work to set up. Since the first PSA tools emerged about two decades ago, they have become easier and easier to set up and configure. Today, virtually all of them have a cloud (hosted) option. I have used several different PSA tools and highly encourage you to invest in one of the top brand names.

Setting up a PSA requires you to do a great deal of thinking about how you structure the flow of work and money through your organization. If you haven't thought about these things, it can be difficult and time-consuming (but also a great business practice!). I have entire chapters on this in *Managed Services in a Month* and in Volume Three of the *Managed Services Operations Manual*.

My preference is for the SMB consulting-specific PSA instead of using a more generic CRM (customer relationship management) tool. The reason is simple: After all the setup labor is done, what you get is extremely SMB consultant focused. And the best packages will integrate more or less seamlessly with QuickBooks and Microsoft Exchange, as well as your RMM.

As you sign more and more service agreements, I believe you will eventually buy a RMM package to provide monitoring and patch management with limited labor. If you want to grow your business with more clients or simply spend less time serving the clients you have, both the RMM software and the PSA package will be very helpful in reaching your goals.

You can get by for quite a while with a home-grown collection of tools. But I believe that you will eventually need to buy the big software packages.

The final section of this chapter takes a more practical turn as we look at cover letters for Service Agreements.

Cover Letters

This section contains cover letters for several scenarios. These cover letters are also included on the accompanying downloads (register your book at www.smbbooks.com) They are:

- Cover letter for a quote for service
- Cover letter to a prospect who has agreed to sign a Service Agreement
- Cover letter to a client to renew an agreement with no changes
- Cover letter to a client to renew an agreement at higher rates
- Cover letter to a client who is adding on a service or upgrading to a higher level of service

Most of the text of these letters is identical. Just the relevant paragraphs are different.

Cover Letter for a Quote for Service

This is the basic letter we use for a quotation cover letter.

```
Text:

Date:      Month, xx, Year
To:        The Customer Client Organization
From:      You the Consultant Your Company

re: Quote for Network Services
```

You should have a lawyer review every agreement or contract you sign.
www.smallbizthoughts.com

Thank you for giving me the opportunity to review your systems and talk to you about your ongoing technical support. As promised, I am enclosing a quote for the work we discussed on your systems. There is one quote for the server and related labor, and another quote for the desktops and related labor.

This proposal includes moving to a Windows Server as your primary server, as we discussed. In addition to the increased reliability for basic file serving, you will have many other features that will make your computer operations faster, more secure, and more reliable.

The quotes assume that we will have a service agreement with a rate of $150 per hour. As you may know, our non-contract labor rate is $165 per hour. Let me know if you have any questions about this.

To begin right away, sign the approval notice and email it to me directly at _____

The most important part of a regular support agreement is the monthly maintenance. During monthly maintenance, we go through a long list of "checks" customized for your machine and network. Generally, these include:

- Check the server logs
- Verify that the backup is working properly
- Check disc fragmentation
- Review Email logs and functionality
- Clean the Tape Drive
- Label end-of-month tape for offsite

storage
- Run Security Analysis
- Run Exchange configuration analysis
- Check Internet Bandwidth
- Check disc space available on the server
- Apply Windows updates as appropriate
- Verify that the virus scanner is working
- and more!

We also offer a fixed-fee service agreement that would include the monthly maintenance, offsite monitoring, and virtually all labor for a flat monthly fee. With offsite monitoring, we provide these major services:

1) Our monitoring system maintains contact with your server and alerts us when there are issues that need attention, such as a failing hard drive, excessive processor usage, or virus activity.

2) We receive alerts, backup reports, and server performance reports from your server on a regular basis. Some of these are daily, some are weekly. Some are every sixty seconds!

3) We connect remotely to your system from time to time and verify that there are no obvious problems.

4) We apply all important updates to Windows, Microsoft Office, and all important software. It is our policy to wait until these updates are stable, so we do not always apply them immediately upon release. But when they are known to be stable, we apply the updates.

Many problems are solved by having decent equipment and applying the latest patches, fixes, and updates. If you've made many tech support calls for hardware or software, you know the first question is, "Are you on the latest service pack?" With SuperStar Tech Support, we make sure your machines are up to spec so that you can spend your energy on your business.

One of the advantages of remote monitoring is that we can easily provide assistance at any time. We can change passwords, unlock accounts, apply updates, and even give desktop support to users by connecting directly into their machines. Another advantage is that you can go on vacation and know that someone is watching the server!

Our Company is dedicated to giving you First-Rate Technical Support. We hire only skilled, trained, and certified professionals. We will work very hard to be responsive when you call, to provide you with good advice, and to keep your systems working as smoothly as possible. You will find our people to be friendly, courteous, and professional.

You can expect us to be honest, competent, and fair. We will provide you with excellent service and your computers will work better every time we work on them. We won't cheat you and we won't leave you "hanging" when you need support.

If you have any questions or problems, please give us a call at 999-555-5555.

Thank you for considering Your Company. We look forward to working with you.

Cover Letter to a Prospect Who Has Agreed to Sign a Service Agreement

Once a prospect agrees to sign an agreement, it is your job to get the agreement in the mail or mail as soon as possible. Even a slight delay can result in a change of heart.

This letter should be very positive and energetic. The tone should be "Welcome to the family." Use phrases such as "we can't wait to get started" or "we're really looking forward to working with you."

Text:

```
Date:      Month, xx, Year
To:        The Customer Client Organization
From:      You the Consultant Your Company

re: Enclosed Service Agreement
```

Thank you for your business, and Welcome to the <u>Your Company</u> "Family."

I have enclosed an "executed" Service Agreement. If you have any questions, please let me know.

Without a service agreement, our hourly rate is $165 per hour. With the service agreement you would receive the discounted rate of $150 per hour. Please sign both copies of the agreement and send one back in the enclosed envelope.

If you wish to move to a managed service agreement, you may do so at any time. There is more information about the managed service agreement on the next page.

Our Company is dedicated to giving you First-Rate Technical Support. We hire only skilled, trained professionals. We will work very hard to be responsive when you call, to provide you with good advice, and to keep your systems working as smoothly as possible. You will find our people to be friendly, courteous, and professional.

You can expect us to be honest, competent, and fair. We will provide you with excellent service and your computers will work better every time we work on them. We won't cheat you and we won't leave you "hanging" when you need support.

All we ask of you is that you recognize that we are a very small business and pay your bills in a timely manner.

If you ever have any questions or problems, please give us a call. We never charge for those little 5-minute phone calls. Call 999-555-5555.

Feel free to use this any time of the day or night. If we don't answer, please leave a message.

— Next Page —

Managed Service - SuperStar Tech Support

Includes remote monitoring of Servers and Workstations, and the ability to remotely update, fix, and support all computers. This service includes all maintenance labor for your computers and systems. Virtually everything related to computer or server support is included. It does not cover hardware failures or special projects not related to maintenance.

SuperStar Tech Support includes the ability to remotely connect to your computers, take control of them, and help you walk through configurations, etc. We will automatically apply service packs, software updates, security updates, and so forth. You can rest assured that your systems are properly configured and maintained.

The most important part of a regular support agreement is the monthly maintenance. During monthly maintenance, we go through a long list of "checks" customized for your machine and network. Generally, these include:

- Check the server logs
- Verify that the backup is working properly
- Check disc fragmentation
- Review Email logs and functionality
- Clean the Tape Drive
- Label end-of-month tape for offsite storage
- Run Security Analysis
- Run Exchange configuration analysis
- Check Internet Bandwidth
- Check disc space available on the server
- Apply Windows updates as appropriate
- Verify that the virus scanner is working

- and more!

Thank you for your business.

Cover Letter to a Client to Renew an Agreement with No Changes

The most important aspect of the basic renewal is to just do it — on time. It's quite embarrassing to send out a note that says "Our service agreement expired three weeks ago. Please sign this so you can continue to receive a preferred rate on labor." I know it's embarrassing because I've done it!

The basic renewal letter points out two important facts – our regular rate is now $ _____ and your rate is not going up. Whenever you have the opportunity to raise rates and don't, make sure you tell the client!

Text:

```
Date:      Month, xx, Year
To:        The Customer Client Organization
From:      You the Consultant Your Company

re: Contract renewal
```

Well, it's time to renew our Service Agreement. As you know, our current Service expires on _____date_____ . I have printed up a new agreement that covers the period ____date____ onward. This agreement is renewed "month to month" until cancelled by either party. There are no changes in rates or minimum charges.

- The basic labor rate is moving to $ ____

per hour. (no change)
• The after-hours rate is twice the regular
rate ($ _____ per hour) (no change)
• There is a one hour minimum for onsite
technical support. (no change)

We never like to raise rates, but sometimes
we have to. As a long-term client under
contract, you have been cushioned from most
rate increases. Your last rate change was .
. . never!

Our current regular rate is $ _____ per hour.
You will still be 10% below that.

Please sign both copies of the agreement and
return one to me. Thank you for your continued
business! We look forward to working together
for a long time to come.

Cover Letter to a Client to Renew an Agreement at Higher Rates

Of course you want to keep your clients and not scare them off with frequent increases, but eventually everything has to go up. If you've gone to $120/hour and still have one or two clients back at $75, then it's time to raise their rates.

We'll look at tracking service agreements later. In general, no one should go too long without a price increase. And if someone's been at the same rate for three years, you should have no problems with a modest increase.

Text:

Date: Month, xx, Year

To: The Customer Client Organization
From: You the Consultant Your Company

re: Contract renewal

Well, it's time to renew our Service Agreement. As you know, our current Service expires on ____(date)__. I have printed up a new agreement that covers the period __(date)__ onward. This agreement is renewed "month to month" until cancelled by either party.
There are a few changes in rates or minimum charges:

- The basic labor rate is moving to $ _____ per hour
- The after-hours rate is twice the regular rate ($ _____ per hour)
- There is a one hour minimum for onsite technical support.

We never like to raise rates, but sometimes we have to. As a long-term client under contract, you have been cushioned from most rate increases. Your last rate change was __year__. Our current regular rate is $ ____ per hour. You will still be 10% below that.

Please sign both copies of the agreement and return one to me.

Cover Letter to a Client Who Is Adding on a Service or Upgrading to a Higher Level of Service

These are pretty easy letters to write and to send. You've got an existing client who has agreed to add some services. This is just a piece of paper that gets the job done.

Text:
< Same as above >

. . .

```
At your request I am also including an
agreement for managed services. With the
managed services agreement you will be billed
a flat fee per month to cover all maintenance
costs for operating systems and software.
Please execute the attachment in addition to
the service agreement.

Please call me if you have any questions.
```

Organizing Letters and Agreements

With your growing catalog of contracts and cover letters, you need to do a little organization — for both the electronic and the paper versions. Perhaps the best way to go about organizing is to think about how you would explain your system to someone else.

For the electronic files, I recommend that you keep all your templates in a single folder, perhaps as read-only documents. Do not get in the habit of editing a template and then saving it to a client folder. You will eventually end up with a template that has been half filled out.

It can be quite embarrassing to send a client something that says:

All General Consulting Hours will be provided at a rate of one hundred twenty dollars ($135.00) per hour.

If a client reads carefully, she'll cross out the $135 and write in $120 to be "consistent." Then you have to explain that you can't really do the deal at $120, no one gets the $120 price anymore, etc. In general, this is an uncomfortable conversation that can be avoided with proofreading and good habits with regard to using templates.

If you keep one folder for your templates, you will also know that the latest version of your agreement is there. This is much better than going to a client folder, copying their service agreement, and pasting it to the new client's folder. If you've made minor changes here and there, you might not have the latest version of your agreement.

The best practice, if you are comfortable with your agreement, is to print up a bunch of them on colored paper, back to back, and have true fill-in spots for the date and name, but do not make the price a fill-in. The price is the price. All new agreements are at the new price.

With such forms, you can actually ink a deal during a sales call and not have to go back to the office, forward a copy of the contract, and give the client an opportunity to back out. I can't tell you how many times the smallest delay has created a larger delay. Even if the client has agreed to sign, putting money in your pocket is not a high priority for them. So pre-printed service agreements go a long way.

As for paper copies, the executed agreements, you also need a procedure for storing them. We find that the easiest place to keep signed agreements is in one big folder. The alternative is to keep agreements in each of the clients' folders. This means that all your agreements are spread out all over the place.

It also means that when you put this year's business files into storage to make room for next year's business files, you have to remember to pick out all those agreements and transfer them to new folders.

Once again, you can do whatever makes the most sense for you. Just make sure that you do the filing and that the agreement is where it's supposed to be. Whatever your system may be, success is found in executing it consistently. Remember, you need to explain this to someone and she needs to be able to find the agreements.

If you don't already have someone to help with administrative duties, then you should be preparing to hire that person. One of the first chores you'll turn over to her is keeping track of all these agreements. If you create a system now that actually works, it will make that transition much easier.

Concluding Thoughts

This chapter covered a lot of the practical side of managing your managed service business. A great deal of it just amounts to pushing one kind of paper or another. Either you're printing up cover letters, or pushing a sale, or filing agreements. These might be boring "housekeeping" chores, but they're also vitally important to your success. Put systems in place to make sure things get taken care of.

We also talked about managing the overall picture. Most consultants find that, when they start out, they have different rates for different clients. Over time, your sanity depends on bringing these rates into line. Clients at the lower end need to join the "norm." This might mean several annual increases, or one big jump. Again, think from the client's perspective. If they've been paying $80 an hour for five years, they know they're overdue for an increase.

Eventually, you'll have 100% of your clients at the same rate. Perhaps then you'll realize that that has been your standard rate for more than three years and you need to change it.

Finally, we discussed the software and tools you will use to actually deliver and manage these services. This book started with a statement about the great changes in our industry. One of the happiest changes is in that our tools are getting better every year — every month, it seems.

A few years ago, you could choose between one or two software packages for running your business and one or two for delivering services. Now the players are changing so fast that I don't want to list them in a book format! (Please do, however, look at the resources section for links to Internet sites that discuss these tools.)

Between the tools included with Microsoft products, those offered by hardware and software vendors, and all-in-one options available, you now have many choices. The sole proprietor can now service and manage many more clients because of the mature tools available. At the same time, she can manage her business more effectively because of these tools.

For larger consultancies, the news is even better. We can manage more employees with less effort, deploy dozens of fixes to thousands of computers in a matter of minutes, and know that all of our products work together and interact with one another.

The bottom line (money, the actual bottom line) can be improved dramatically if you use the right tools, have a system that works, and use it consistently. If you don't have a system that works for your business, then signing agreements will do you some good, but you won't be getting the full benefit of being a managed service provider.

"Live together like brothers and do business like strangers."
— Arab Proverb

Conclusions

"A storm is gathering around us. We can ignore the change, we can fight the change, but we can't stop the change. There is one more alternative – we can make our businesses evolve to take advantage of the change."
— Chapter One

Did you notice what happened as you read this book? In Chapter One I warned about the changes to come and gave some speculations about where our industry is going. By the end of Chapter Nine I felt comfortable referring to your small business as a managed service provider.

Small business consultants are very aware of the fact that it's easy to get into this business. You order some business cards that say "Computer Consultant" and, Poof! You're a computer consultant.

That doesn't mean you're a good consultant, or a good technician, or a successful business person. It just means you bought some business cards.

The bane of our existence is the amateur techno-goober who gets into this business and does a horrible job. That guy makes my life very difficult sometimes. I hate going on sales calls and having to tell the client "Your last consultant was a dishonest, incompetent imposter. We're different!" I would hate to be the client who has to listen to that.

In my humble opinion, the best thing we can do for our industry is to be professionals, to act like professionals, and to treat our clients in a professional manner. Among other things, that means we need to stand behind our work and treat that client relationship with respect.

I believe service agreements are the key to this process. I think it's no coincidence that managed services are being offered by virtually all of the larger players. Modern tools have made it possible for us to offer these services at a reasonable price. Of course, that's not a good enough reason.

Being a Managed Service Provider (MSP) allows you to provide a higher level of service to your clients. You should use this as a selling point! The kid down the street who's going to disappear from the business in six months will not be signing any service agreements. He also won't promise a two hour response time, invest in a ticketing system, or provide any of the latest remote support options.

Here are a final few tips to help you get started along the road to becoming a Managed Service Provider.

1. Let Other People Help You

In addition to basic feedback mechanisms, there are several ways that other people can help you. The most obvious of these is to get some administrative support. For the small, small sum of $120 per week, you can hire someone to help you out with filing and paperwork, taxes, invoicing, mailings, and more. That's eight hours at $15 per hour. If nothing else, put this person on the phone collecting past-due invoices.

Here's another easy tip – bring in someone you trust and explain to that person how you run your business. Explain your pricing structure, your billing process, and how you determine what's billable and what's not. This could be your financial advisor, your spouse, another consultant, or just a friend who is willing to listen and ask questions.

If you can't explain your system to someone else, you probably haven't really created it, you just let it grow on its own. Which leads to #2.

2. Define Yourself

Look through Chapters One and Two again. How do you want to run your business? Are you where you intended to be, or are you just where you ended up? More importantly, how will your business change in the next five years? Where will you be? Where do you want to be?

The next five years are going to pass, and your business is going to change. If you have some plans and some ideas and you know where you want to go, you will probably get there. If you just bounce around like a pinball, then you don't have any idea where you'll be in five years. Statistically, we know that eighty percent of all businesses fail in the first five years. Of those that survive, eighty percent fail in the next five years.

Sadly, that means there's no learning curve. And if you choose to be the pinball, allowing your business to be pushed around by circumstances, then chances are very good that you'll be down the hole. Gone.

Begin planning today. In the resources section I recommend a book that every SMB consultant must read: *The E-Myth Revisited* by Michael Gerber. Start reading there. Then begin working on your business as well as working in your business.

3. Draft Your Managed Service Agreement
(and have it reviewed by a lawyer)

After you know who you want to be and what kind of relationships you want to have with your clients, begin drafting an agreement that will get you there. Please don't just copy the text out of this book! This really is just a starting place.

Every aspect of your business should be your own and not someone else's. Which leads to . . .

4. Build Business Processes That Reflect Your Way of Doing Business

Over and over again in this book I've stated that you need to do things your way. You need to be comfortable with the business formation, with the target clients, with the service agreements, with the pricing, with the marketing, and so forth.

When I get together with consultants who are eager to grow, I am often worried that they are too eager to take someone else's successful system and adopt it as their own. One of the greatest pleasures of owning your own business is building something that's really your own.

We all want to be successful and make money. But if you find yourself building a business you don't like, selling products and services you don't feel confident to deliver, and being tied to an "outside" software package that forces you to do things its way, that can lead to burn-out very quickly. Don't build a business you hate just because the pieces made sense for someone else.

Use the tools (including service agreements) when they fit and feel comfortable for your business. Where they don't fit, don't use them. When they're not comfortable, don't use them. Remember, these are promises. Don't make promises you don't feel comfortable with.

5. Get Help: Legal, Financial, and Insurance

In addition to your $120 per week miracle assistant, I highly recommend that you find and use good people in the areas of law, finances, and insurance.

These people will help you stay organized in addition to making sure your business is safe and properly constituted. Just as your clients hire you for professional technical services, you should hire professionals for these important areas.

6. Pick Out the Tools You Will Use to Manage and Deliver Services

To a technician, this is the one piece of advice that sounds fun. And it is! Whether you start with a roll-your-own solution, or buy a fancy software package, you will need some tools to provide the level of service clients expect in the twenty-first century.

Whatever you do, make a plan and make it your own. Remember, the next five years are going to pass whether you plan them or not.

You should have a lawyer review every agreement or contract you sign.
www.smallbizthoughts.com

"The best investment is in the tools of one's own trade."
—Benjamin Franklin

Resources

Resources from This Book

- Downloads for the book
www.smbbooks.com

- The American Arbitration Association
http://www.adr.org/

- The American Bar Association
www.abanet.org

- American Institute of Certified Public Accountants
www.aicpa.org

- The ASCII Group
www.ascii.com

- CPA Directory
www.cpadirectory.com

- Docusign
www.docusign.com

- Internal Revenue Service
http://www.irs.gov

- IRS's "Tax Topic 751"
https://www.irs.gov/taxtopics/tc751.html

You should have a lawyer review every agreement or contract you sign.
www.smallbizthoughts.com

- IT Insure
www.itinsure.com

- LexisNexis
www.lawyers.com

- Find a mediator
www.mediate.com

- National Association of Enrolled Agents
www.naea.org

- The National Bar Association
www.nationalbar.org

- Robin Robins' Technology Marketing Toolkit
www.technologymarketingtoolkit.com

- Sadler & Company, Inc.
www.insurancefortechs.com

- Servers Alive
www.woodstone.nu/salive

- State-level CPA associations

- State-level Enrolled Agents associations

- Tech Shield
www.techshield.com

- Techinsurance
www.techinsurance.com

- Your State Bar Association

When you're ready to start growing your consulting practice at a faster pace, here are some resources that will give you a quick boost. I can't list every great resource out there, but here are a few prize picks. The list includes books and web sites. After that you'll find a list of additional resources available in the downloads that accompany this book.

Books

If you're interested in a much longer list of books that I recommend, please see the Books link at www.RelaxFocusSucceed.com. I have only listed a few books here that I think are extremely valuable to your SMB Consulting practice.

The first is listed out of order because it's the first book you should read or re-read. After that, books are in order by author's last name.

Must-read books for all SMB consultants:

The E-Myth Revisited by Michael Gerber

If I could make every small business owner read just one book, this would be it. Really. It's a quick read, but has some very powerful messages about how to make your business (and your life) successful without working yourself to death.

Other excellent books:

SMB Consulting Best Practices by Harry Brelsford
Now a bit old, but this is still an excellent book for SMB consultants. Harry refers to it as your MBA in a book. This book provides a wonderful framework for looking at your business and your business relationships.

The Power of Focus by Jack Canfield, Leslie Hewitt, Mark Victor Hansen

Perhaps nothing is more difficult for small business owners than to stay focused on the things that matter. Most of us have goals, and some of us have well-defined goals. And yet our industry is typified by people who work until midnight. It doesn't have to be that way.

The Business Startup Checklist and Planning Guide by Stephanie Chandler

This is a great book for those of you just starting out in the business. But it is also good for people who are sole proprietors and who may have skipped some of the important early steps when they set up their business.

The Seven Spiritual Laws of Success: A Practical Guide to the Fulfillment of Your Dreams by Deepak Chopra

Chopra has many excellent books, but this one really stands out to me because of the focus on fulfilling your dreams. Having dreams and fulfilling dreams is more fun and exciting than having goals and reaching goals.

First Things First by Stephen R. Covey, A. Roger Merrill, and Rebecca R. Merrill

For those interested in working on the business side of your business, this is a great book. In particular, if you want to know why you're working so hard and not getting ahead, this book is for you.

The E-Myth Contractor by Michael Gerber

Gerber revisits the E-Myth book and applies lessons for the world of contracted labor. Not 100% on target with the computer business, but still very good.

Who Moved My Cheese? An Amazing Way to Deal With Change in Your Work and in Your Life by Spencer Johnson

You've certainly heard about this book, even if you haven't read it. It really is good and you really should read it. Much better than Who Stole My Fish or whatever his later book was. If your organization is going through change—such as implementing a managed services offering—then you need to read this book.

Answers for Computer Contractors by Janet Ruhl
Old but good. Ruhl has written several books for technical consultants. All of them are pretty good. I like this one because of the Q&A approach. Really great, sound advice for getting your new consulting business up and running.

Authors

There are some authors who produce so many great books that it is a disservice to just mention one book. The following authors are truly great and inspiring. Any book by these people is a safe bet. Just pick one up and start reading. You won't be disappointed. They all also produce audio programs.

- Jim Collins
- Wayne Dyer
- Seth Godin
- Brian Tracy
- Zig Ziglar

Web Sites

As with books, there are too many web sites to list. Some of my favorites, which I have bookmarked and which I visit every day, are not listed. I believe these few web sites are the cream of the crop for people who want to focus on growing their SMB consulting practice.

Sites are in alpha order by title.

ChannelPro Network
http://www.channelpronetwork.com/

Publisher of ChannelPro magazine and producer of great managed service-focused events. A great aggregator of news and information.

Computer Information Agency (Robert Crane's blog)
http://blog.ciaops.com/
Robert gives great advice for SMB IT professionals and is always pushing new, important technologies you should be using and selling to your clients. Robert is a long-time contributor to our community from Australia.

Great Little Book Publishing / SMB Books
http://www.greatlittlebook.com or http://www.smbbooks.com
Hey, if you don't toot your own horn, who's gonna do it for you? We try to publish good books for the SMB consulting audience. Talk about a niche.

Relax Focus Succeed
http://www.relaxfocussucceed.com
This is my "other" job. In addition to more than fifty articles to help you build your success, we have a year's worth of newsletters online. You can also subscribe to our free monthly email newsletter.

SMB Nation
http://www.smbnation.com
Home of Harry Brelsford, the absolute guru of the SMB IT consulting world. Find out about upcoming conferences, new book releases, Harry's newsletter, one-day summits, and much more.

Tubblog (Richard Tubb's blog)
http://www.tubblog.co.uk/
Former MSP and business coach Richard Tubb gives you news and commentary from the UK. He is widely traveled and always brings interesting information.

Windows Server Blog
https://blogs.technet.microsoft.com/windowsserver
The official Microsoft bog for Windows Server.

What's in the Downloads?

The Download Content that comes with this book has all of the agreement language, cover letters, and Excel spreadsheets discussed in the text, plus much more.

If you don't have Microsoft Word, or your version is too old to open these files, please send the book back to us for a full rebate as you are obviously in the wrong business.

The primary links are:

- Sample Agreements
- Spreadsheets
- The various spreadsheets for speculating on prices, and for keeping track of service agreements.
- Cover Letters
- The various cover letters discussed in Chapter Eight.
- IRS Goodies
- Some tax code snippets, links, and .PDF files from the IRS web site.
- Links and Resources
- Places to go, books to read, etc.

Karl's Email

Here's one more resource: My email address is

karlp@smallbizthoughts.com

Please send me your feedback and questions.
Thank you once again for purchasing this book.

Acknowledgements

Sometimes an author will say, "This book would not have been written without the encouragement of . . .." Well, let me tell you right up front, THIS book would not have been written without a kick in the butt from Harry Brelsford.

Harry called me one day way back in 2006 and said, "You know what's missing in our industry? A book on contracts. Can you do that?" I told him I'd try. So here it is. Thank you, Harry.

I also thank Harry for all of his support and encouragement regarding *The Network Documentation Workbook, The SAN Primer for SMB*, and for many projects over the years. And I thank Nancy Williams for being instrumental in putting my first book in front of Harry's eyes when he was super busy. She's been a great friend ever since.

Of course thanks go out to Yvonne Betancourt for her layout work, and to Sally Galli for the cover design. Since I squished this book in between tours across Europe, the U.S., and Australia, I appreciate everyone's flexibility.

Thanks also to Hank Roitman, Enrolled Agent, for advice on a couple of chapters.

If I forgot anyone, please forgive me.

— Karl W. Palachuk
Sacramento, CA

About the Author

Karl W. Palachuk is the author of seventeen books. He has built and sold two successful Managed Service businesses in Sacramento, CA. He is the founder and past president of the Sacramento SMB IT Pro User Group. His other books include *Managed Services in a Month* and the 4-volume set **The Managed Services Operations Manual.**

Karl has been a featured speaker at conferences and seminars over the last fifteen years. He is a Microsoft Certified Systems Engineer with a Bachelor's Degree from Gonzaga University and a Master's Degree from The University of Michigan. In 2005 Karl was one of the first Microsoft partners to become a Small Business Specialist. He has also been a Microsoft Hands-On Lab instructor.

In addition to all this technical work, Karl writes and operates a motivational training company called Relax Focus Succeed. You can read about that at www.relaxfocussucceed.com, and sign up for the newsletter.

Karl's popular blog, "Small Biz Thoughts" can be found at blog. smallbizthoughts.com.

Karl lives in Sacramento, CA.

Keeping Up with Karl

The Blog
Check out Karl's Blog at
http://blog.smallbizthoughts.com
Covers information on Karl's seminars as well as tips on running a small technology consulting practice.
Available on RSS.

The Mailing List
Go to www.smallbizthoughts.com to sign up for Karl's low-volume email list. You'll get updates on book projects, Karl's schedule, comments on the feedback we get from books, and lots of business tips and standard operating procedures.

Please also take a minute to connect with me on Twitter, Facebook, Google+, and LinkedIn. Just search for "KarlPalachuk" or Karl Palachuk on any of those services.

Also look for my SOP (standard operating procedure) videos on YouTube. By the time this book is printed, I should have about 350 videos at www.youtube.com/smallbizthoughts.

Email
Stay in touch. Email Karl at: karlp@SmallBizThoughts.com.

Other Books by Karl W. Palachuk:

Managed Services in a Month – Build a Successful IT Services Business in 30 Days – 3rd Edition

Managed Services Operations Manuel: Standard Operating Procedures for Computer Consultants and Managed Service Providers – 4 volume set
- Volume 1: Front Office Mastery
- Volume 2: Employees & Internal Processes
- Volume 3: Running the Service Department
- Volume 4: Support & Service Delivery

Relax Focus Succeed – Balance Your Personal and Professional Lives and Be More Successful at Both – Revised Edition

Project Management in Small Business – How to Deliver Successful, Profitable Projects on Time with Your Small Business Clients (with Dana Goulston, PMP)

The Network Documentation Workbook

The Network Migration Workbook (with Manuel Palachuk)

Find all of these and more at **www.SMBBooks.com**!

The Managed Services Operations Manual – 4 vol. set
by Karl W. Palachuk

Standard Operating Procedures for Computer Consultants and Managed Service Providers

Every computer consultant, every managed service provider, every technical consulting company - every successful business - needs SOPs!

When you document your processes and procedures, you design a way for your company to have repeatable success. And as you fine-tune those processes and procedures, you become more successful, more efficient, and more profitable.

Relax Focus Succeed®
Balance Your Personal and Professional Lives and Become More
Successful in Both
by Karl W. Palachuk

The premise of this book is simple but powerful: The fundamental keys to success are focus, hard work, and balance. Too often, the advice we receive gives plenty of attention to focus and hard work, but very little to balance.

You should have a lawyer review every agreement or contract you sign.
www.smallbizthoughts.com

You should have a lawyer review every agreement or contract you sign.

You should have a lawyer review every agreement or contract you sign.
www.smallbizthoughts.com

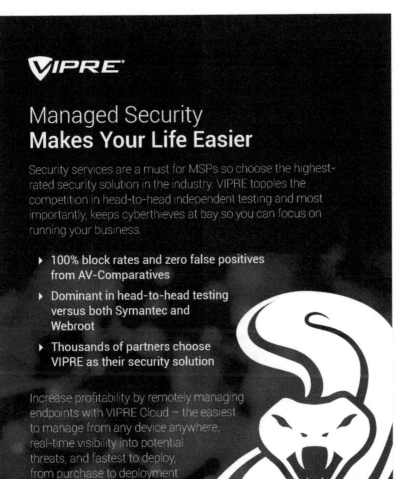

You should have a lawyer review every agreement or contract you sign.
www.smallbizthoughts.com

CPSIA information can be obtained
at www.ICGtesting.com
Printed in the USA
BVHW052231300721
612910BV00008B/759